Advances in Intelligent Systems and Computing

Volume 1332

The series "Advances in Intelligent Systems and Computing" contains publications on theory, applications, and design methods of Intelligent Systems and Intelligent Computing. Virtually all disciplines such as engineering, natural sciences, computer and information science, ICT, economics, business, e-commerce, environment, healthcare, life science are covered. The list of topics spans all the areas of modern intelligent systems and computing such as: computational intelligence, soft computing including neural networks, fuzzy systems, evolutionary computing and the fusion of these paradigms, social intelligence, ambient intelligence, computational neuroscience, artificial life, virtual worlds and society, cognitive science and systems, Perception and Vision, DNA and immune based systems, self-organizing and adaptive systems, e-Learning and teaching, human-centered and human-centric computing, recommender systems, intelligent control, robotics and mechatronics including human-machine teaming, knowledge-based paradigms, learning paradigms, machine ethics, intelligent data analysis, knowledge management, intelligent agents, intelligent decision making and support, intelligent network security, trust management, interactive entertainment, Web intelligence and multimedia.

The publications within "Advances in Intelligent Systems and Computing" are primarily proceedings of important conferences, symposia and congresses. They cover significant recent developments in the field, both of a foundational and applicable character. An important characteristic feature of the series is the short publication time and world-wide distribution. This permits a rapid and broad dissemination of research results.

Indexed by SCOPUS, DBLP, EI Compendex, INSPEC, WTI Frankfurt eG, zbMATH, Japanese Science and Technology Agency (JST), SCImago.

All books published in the series are submitted for consideration in Web of Science.

More information about this series at http://www.springer.com/series/11156

Salvatore Greco · Mario F. Pavone ·
El-Ghazali Talbi · Daniele Vigo
Editors

Metaheuristics for Combinatorial Optimization

Springer

Editors
Salvatore Greco
Department of Economics and Business
University of Catania
Catania, Italy

El-Ghazali Talbi
Laboratoire Cristal
University of Lille and Inria
Lille, France

Mario F. Pavone
Department of Mathematics
and Computer Science
University of Catania
Catania, Italy

Daniele Vigo
Department of Electrical, Electronics,
and Information Engineering and CIRI-ICT
University of Bologna
Bologna, Italy

ISSN 2194-5357 ISSN 2194-5365 (electronic)
Advances in Intelligent Systems and Computing
ISBN 978-3-030-68519-5 ISBN 978-3-030-68520-1 (eBook)
https://doi.org/10.1007/978-3-030-68520-1

This Springer imprint is published by the registered company Springer Nature Switzerland AG
The registered company address is: Gewerbestrasse 11, 6330 Cham, Switzerland

Editorial

It is well known, and it is easy to prove that any problem can be formulated and tackled as an optimization problem since solving it means basically making decisions. Every day each of us continually makes decisions during own daily activities, from simple and automatic ones (e.g., choose a food or dress to wear), to more challenging and complex ones (e.g., in which stock market to invest), and, further, such decisions must be taken quickly and effectively. Making decisions, mean choosing an action or an option from several alternatives according to an expected utility, that is, select among all available options the one that optimize a given goal, such as for example the classical goal of maximizing profits and minimizing costs. However, making decision is very often challenging and complex due to uncertainty, and/or the large number of information to be handled; it then follows that many real-world problems are difficult to solve through exact methods and within reasonable times. Approximate algorithms are the main and often unique alternative in solving such problems, thanks to their ability in efficiently exploring large search spaces by reducing their effective sizes.

The metaheuristics are a special class of approximate algorithms, successfully applied on many complex problems such as in economics, industry and sciences in general. They are considered as upper-level heuristic methodologies, which combine classical heuristic rules with random searches in order to solve large and hard optimization problems. However, unlike the exact methods, the metaheuristics find near optimal solutions and in a quick way. Metaheuristic search methods are basically used on those problems, where (1) there is no idea on what the optimal solution is; (2) neither how to go about finding it; and, primarily, (3) when no brute-force search can be computationally performed due to the too large solutions space.

MESS 2018 (*Metaheuristics Summer School*) was the first edition of the international summer school entirely dedicated to the metaheuristics' research field, aimed to students and young researchers, academics and industrial professionals, in order to provide an overview on the several metaheuristics techniques and an in-depth analysis of the state of the art. The main aim of this edition was to inspect and analyze all metaheuristics from its designing to its implementation, as well as analyze the modern heuristic methods for search and optimization problems, and the classical exact optimization methods seen from in metaheuristics context. Among the several activities offered by the school for its participants, the *Metaheuristics Competition* is the most challenging and interesting one, where each student, during the school, must individually develop a metaheuristic to solve a given combinatorial optimization problem, designed and proposed by the competition chairs.

The Traveling Salesman Problem (TSP) is one of the most studied and well-known combinatorial optimization problem whose goal is to find a routing in a network of minimal distance/cost that starting from a home location (starting city) visits all other cities with the constraint that each city must be visited exactly once. The distance or cost depends on the cities visiting order and, therefore, this means to find an optimal ordering of the cities. TSP is a *NP*–complete problem, and, although significant progress has been made for solving it by exact algorithms, there are still several large instances hard to be solved within reasonable times using exact methods. It follows that approximate heuristic methods have had increasing consideration in tackling this problem, thanks to their ability to produce solutions as close as possible to the optimal one.

Depending on the nature of the network, and then on its cost matrix, the TSP can be divided into two classes: Symmetric TSP (STSP) when the graph is undirected, and Asymmetric TSP (ATSP) when instead it is directed. Note that any undirected graph can be seen as a directed graph, just only duplicate the edges (forward and backward), it follows that STSP is a special case of the ATSP. Although the features, structure and constraints related to the TSP make it very suitable mainly for route planning problems, it finds instead application in many and different areas, such as for instance, (*i*) machine sequencing and scheduling problem; (*ii*) cellular manufacturing; (*iii*) frequency assignment; and (*iv*) multiple sequence alignment, just for citing a few. Moreover, the TSP model finds applicability also in data analysis in psychology [10], X-Ray crystallography [2], overhauling gas turbine engines [15], warehouse order-picking problems [16], and wall paper cutting [6]. Thanks to that the TSP is nowadays the most challenging and more studied combinatorial optimization problem.

As a result of these different applications, several variants of the classic TSP have been designed and inspected, each of them originating from various real-life problems, such as for instance:

- *TSP with Multiple Visit*, where each node is visited at least once;
- *Time-Dependent TSP* [3,18], where the weight on the edge from i to j is the cost of crossing the edge in the time period t;
- *Period TSP* [4, 14], where a k–day planning period is considered;
- *Minimum Latency Problem* [17] (also known as *Traveling Repairman Problem*), whose goal is to minimize the sum of path lengths to all vertices;
- *TSP with Time Windows* [1,12], where each edge can be crossed only in a given time interval (if the salesman arrives at given node v_i before the crossing time for reaching v_i, then he will have to wait);

- *Moving Target TSP* [18], whose goal is to intercept in the fastest possible time the set of nodes (targets) that move with constant velocities;
- *TSP with Pickup and Delivery* [5, 9], where exist pickup and delivery vertices and the goal is to determine the minimum cost tour such that any pickup vertex is visited before its corresponding delivery vertex.

More details on the TSP problem and its variants, included formulations, applications and algorithms developed, can be found in [7].

Among all the potential variants of the TSP, a challenging and interesting but likely least studied is the one that ask to find the tour whose cost is as equitable (uniform) as possible, which is the case where the required goal is to minimize the difference between the most costly and least costly tour. This problem is well known as *Balanced Traveling Salesman Problem* (BTSP) [11]. A classic example of balanced optimization, described and reported in [13], is the tour design problem of a travel agency: given n local agencies in n different countries in Europe, the goal of the American travel agency is propose to their customers a set of trips whose lengths are as equal as possible. Each trip is proposed by each local agency, and one trip is offered to each country. Formally, BTSP can be defined as follow: let $G = (V, E)$ be a complete graph, c_{ij} the cost associated to the edge $(i,j) \in E$, and let $\Pi(G)$ be the collection of all Hamiltonian cycles in G; the goal of BTSP is to

$$minimize\{max_{(i,j)\in H}\ c_{ij} - min_{(i,j)\in H}\ c_{ij}\}, \tag{1}$$

with $H \in \Pi(G)$.

Thus, the BTSP has been considered as the test-problem for the Metaheuristics Competition at MESS 2018 in order to design and implement a metaheuristic by the students for solving this problem. In particular, given an undirected graph G, and considering the costs in the edges are positive or negative values, the goal of the proposed test-problem is to find a tour with a cost as close as possible to zero. Such version is called *Cost-Balanced TSP*. For the competition, more than 27 instances have been produced, from medium to more complex ones and with different number of nodes (from small to large). For each instance, each metaheuristic developed has been evaluated with respect the best solution found by the proposed approach and ranked according to the total score obtained. Specifically, a score is assigned to each competitor if the submitted solution for a given instance is in the range (0,1): maximum score 1, for instance, if it is the best submitted solution so far. The goal of the competition is then to get the highest sum of the scores: maximum total score is 27. In this, special issue is presented and described the best top four metaheuristics, respectively: (1) *"Mixed Integer Programming formulations for the Balanced Traveling Salesman Problem with a lexicographic objective"* by Libralesso; (2) *"A Memetic Random Key Algorithm for the Balanced Traveling Salesman Problem"* by Aslan; (3) *"A Variable Neighborhood Search Algorithm for Cost-Balanced Traveling Salesman Problem"* by Akbay and Kalayci; and (4) *"Adaptive Iterated Local Search with Random Restarts for the Balanced Traveling Salesman Problem"* by Pierotti, Ferretti, Pozzi and van Essen.

Finally, the editors would like to thank all the competitors of MESS 2018 Metaheuristics Competition for their interesting and novelty metaheuristic approaches developed, and the competitors of the top four in the ranking for their high manuscripts quality presented and included in this special issue. A warm thanks is also given to the editorial manager staff for their professionalism, patience and important support. A final big thank is for the series editor, Prof. Janusz Kacprzyk, for encouraging and accepting this special issue and for his valuable suggestions.

Salvatore Greco
Mario F. Pavone
El-Ghazali Talbi
Daniele Vigo

References

1. N. Ascheuer, M. Fischetti and M. Grötschel: Solving the Asymmetric Travelling Salesman Problem with time windows by branch-and-cut. *Mathematical Programming*, vol. 90, pp. 475–506, 2001.
2. R. G. Bland and D. F. Shallcross: Large traveling salesman problems arising from experiments in X-ray crystallography: A preliminary report on computation. *Operations Research Letters*, vol. 8, no. 3, pp. 125–128, 1989.
3. V. Cacchiani, C. Contreras-Bolton and P. Toth: Models and Algorithms for the Traveling Salesman Problem with Time-Dependent Service Times. *European Journal of Operational Research*, vol. 283, no. 3, pp. 825–843, 2020.
4. IM. Chao, B. L. Golden and E. A. Wasi: A New Heuristic for the Period Traveling Salesman Problem. *Computers & Operations Research*, vol. 22, no. 5, pp. 553–565, 1995.
5. I. Dumitrescu, S. Ropke, J. F. Cordeau and G. Laporte: The traveling salesman problem with pickup and delivery: polyhedral results and a branch-and-cut algorithm. *Mathematical Programming*, vol. 121, no. 2, pp. 269–305, 2010.
6. R. S. Garfinkel: Minimizing wallpaper waste part 1: a class of traveling salesman problems. *Operations Research*, vol. 25, no. 5, pp. 741–751, 1977.
7. G. Gutin and A. P. Punnen: *The Traveling Salesman Problem and Its Variations*. Part of the Combinatorial Optimization book series (COOP, vol. 12), Kluwer Academic Publishers, 2004.
8. C. S. Helvig, G. Robins and A. Zelikovsky: The Moving-Target Traveling Salesman Problem. *Journal of Algorithms*, vol. 49, no. 1, pp. 153–174, 2003
9. H. Hernández-Pérez and J. J Salazar-González: A branch-and-cut algorithm for a traveling salesman problem with pickup and delivery. *Discrete Applied Mathematics*, vol. 145, no. 1, pp. 126–139, 2004.
10. L. J. Hubert and F. B. Baker: Applications of combinatorial programming to data analysis: the traveling salesman and related problem. *Psychometrika*, vol. 43, pp. 81–91, 1978.
11. J. Larusic, and A. P. Punnen: The Balanced Traveling Salesman Problem. *Computers & Operations Research*, vol. 38, pp. 868–875, 2011.
12. M. López-Ibáñez, C. Blum, J. W. Ohlmann and B. W. Thomas: The Travelling Salesman Problem with Time Windows: Adapting Algorithms from Travel-Time to Makespan Optimization. *Applied Soft Computing*, vol. 13, no. 9, pp. 3806–3815, 2013.
13. S. Martello, W. R. Pulleyblank, P. Toth, and D. de Werra: Balanced Optimization Problems. *Operations Research letters*, vol. 3, no. 5, pp. 275–278, 1984.
14. G. Paletta: The Period Traveling Salesman Problem: a New Heuristic Algorithm. *Computers & Operations Research*, vol. 29, no. 10, pp. 1343–1352, 2002.

15. R. D. Plante: The nozzle guide vane problem. *Operations Research*, vol. 36, no. 1, pp. 18–33, 1988.
16. H. D Ratliff and A. S. Rosenthal: Order-picking in a rectangular warehouse: a solvable case of the traveling salesman problem. *Operations Research*, vol. 31, no. 3. pp. 507–521, 1983.
17. M. M. Silva, A. Subramanian, T. Vidal and L. Satoru Ochi: A simple and effective metaheuristic for the Minimum Latency Problem. *European Journal of Operational Research*, vol. 221, no. 3, pp. 513–520, 2012.
18. D. Tas, M. Gendreau, O. Jabali and G. Laporte: The Traveling Salesman Problem with Time-Dependent Service Times. *European Journal of Operational Research*, vol. 248, no. 2, pp. 372–383, 2016.

The original version of the book was revised: The volume number has been amended. The correction to the book is available at https://doi.org/10.1007/978-3-030-68520-1_5

Contents

Mixed Integer Programming Formulations for the Balanced Traveling Salesman Problem with a Lexicographic Objective

Luc Libralesso(✉)

Univ. Grenoble Alpes, CNRS, Grenoble INP, G-SCOP, 38000 Grenoble, France
luc.libralesso@grenoble-inp.fr

Abstract. This paper presents a Mixed Integer Program to solve the Balanced TSP. It exploits the underlying structure of the instances and is able to find optimal solutions for all the instances provided in the Metaheuristics Summer School competition. We study the efficiency of this new model on several variants of the Balanced TSP. The proposed method was ranked first in the MESS18 Metaheuristic competition among 9 submissions.

Instances and ranking: http://195.201.24.233/mess2018/home.html..

Source code: https://gitlab.com/librallu/balancedtspcode.

Keywords: Balanced TSP · MIP · Metaheuristics

1 Introduction, Problem Description, and Preliminary Results

This paper presents a MIP formulation for a kind of optimization problem that involves the minimization of an absolute lexicographic value. It obtains optimality proofs for all instances of the Metaheuristic Summer School 2018 competition (http://195.201.24.233/mess2018/home.html) in a (relatively) short amount of time.

Consider a graph $G = (V, E)$, and a weight function $w : E \to \mathbb{Z}$. We want to find a Hamiltonian tour T that minimizes $|\sum_{e \in T} w_e|$.

We can formulate this problem as follows:

$$\min \quad z$$

$$\text{subject to}$$

$$\sum_{j \in V,\, ij \in E} x_{ij} = 2 \qquad \forall i \in V \tag{1}$$

$$\sum_{i,j \in S,\, ij \in E} x_{ij} \leq |S| - 1 \qquad \forall S \subset V, S \neq \emptyset \tag{2}$$

S. Greco et al. (Eds.): MESS 2018, AISC 1332, pp. 1–15, 2021.
https://doi.org/10.1007/978-3-030-68520-1_1

$$z \geq \sum_{e \in E} x_e w_e \tag{3}$$

$$z \geq -\sum_{e \in E} x_e w_e \tag{4}$$

$$x_e \in \{0,1\} \qquad \forall e \in E$$

$$z \in \mathbb{Z}$$

Constraints (1) and (2) are tour constraints. Constraint (1) guarantees the degree of vertices to be 2 in the tour. Constraint (2) guarantees no subtour. Constraints (3) and (4) make z be the absolute value of the sum of edges in the tour.

If used as is, this formulation gives poor results (see Table 1). A significant improvement can be achieved by taking into account some properties of the instances presented in the competition.

1.1 A Better MIP Formulation

The MESS2018 competition[1] instances represent sparse graphs. Indeed, an instance of size n contains $4 \cdot n$ edges. It implies that the resulting MIP models contain a (relatively) small number of variables. Moreover, it turns out that each weight w can be written as $w = \overline{w} \cdot 10^5 + \underline{w}$, with $\overline{w} \in [0, 10]$ and $\underline{w} \in [0, 100]$. These weights are lexicographic since a solution that does not optimize the $\overline{w}$ part will be worse than a solution that optimizes $\overline{w}$ (unless there are many edges in the instance). Using this property, we can preprocess the input to decompose weights and derive a new model that introduces 3 objectives. One for the sum of $\overline{w}$, one for the sum of $\underline{w}$ and the last one for the sum of w.

$$\min \quad z$$

subject to

$$\sum_{j \in V,\, ij \in E} x_{ij} = 2 \qquad \forall i \in V \tag{5}$$

$$\sum_{i,j \in S,\, ij \in E} x_{ij} \leq |S| - 1 \qquad \forall S \subset V, S \neq \emptyset \tag{6}$$

$$\overline{z} = \sum_{e \in E} x_e \overline{w_e} \tag{7}$$

$$\underline{z} = \sum_{e \in E} x_e \underline{w_e} \tag{8}$$

$$z \geq \overline{z} \cdot 10^5 + \underline{z} \tag{9}$$

$$z \geq -(\overline{z} \cdot 10^5 + \underline{z}) \tag{10}$$

$$x_e \in \{0,1\} \qquad \forall e \in E$$

$$z, \overline{z}, \underline{z} \in \mathbb{Z}$$

---

[1] http://195.201.24.233/mess2018/home.html.

Constraint (5) guarantees the degree of vertices to be 2 in the tour. Constraint (6) eliminates subtours. Constraints (7) and (8) define $\overline{z}$ (resp. $\underline{z}$) to be equal to the weighted sum of selected edges. Constraints (9) and (10) define z to be equal to the absolute value of the original weighted sum of edges.

In the remaining of this document, we call this model the *lifted MIP model.* We call the original model the *standard MIP model.*

1.2 Numerical Results

Results were obtained on an Intel(R) Core(TM) i5-3470 CPU @ 3.20 GHz core with 8 GB RAM and Linux Mint 11. The MIP models were implemented using Gurobi 8.1 and python2.7. Table 1 reports optimal values, and computation times and gaps. In the dataset, instances are denoted by their size. For example, instance 0500 has 500 vertices. Column *optimal obj* shows the value obtained by the *lifted MIP model.* Column *time to opt* reports the time needed by the *lifted MIP model* to find the optimality proof. Finally columns *lb standard, ub standard, gap standard* ($gap = \frac{ub-lb}{ub}$) report bounds obtained by the *standard MIP model* within 30 s. We note that no more time is needed to observe a clear difference in performance between the two models.

As the results show, lifting the variable space in the second MIP has a dramatic effect on performance. Indeed, it allows the solver to branch in order to optimize first the $\overline{w}$ part of the problem, consequently improving the search speed. During the competition, I tried many matheuristic approaches (mostly local branching like). At some point, I noticed that the lifted approach was able to solve optimally the competition instances in a few seconds. It would be interesting to compare the lifted approach with other meta-heuristics (like local branching) on bigger instances (*i.e.* 5.000, 10.000 *etc.*).

In Appendix A, Figs. 1, 2, 3, 4, 5 and 6 show the convergence curves of the *lifted MIP formulation.* We observe very different behaviours on the convergence curves. In the btsp0100 and btsp0250 instances, the upper bound converges quickly and the lower bound is found in the later stages of the resolution. In the btsp0500 instance, the dual bound converges quickly to some good value. However, there is still an important gap to optimality that is closed later during the resolution. In the btsp0700 instance, both best solution found and dual bound improve at the same time to some very small gap (less than 1%). Finally, we observe on btsp1000 and btsp1500 instances that finding a feasible solution takes time (sometimes half the resolution time) and the lower bound of good quality is found in the later stages of the resolution.

In this section, we investigated the lifting process. It allows a dramatic performance increase. This allows us to find optimal solutions for all instances of the benchmark. In the next section, we investigate this impact on different variants of the problem (namely assignment variant, subset variant, balanced spanning tree and regular TSP).

Table 1. Comparison of the standard and lifted formulations for the balanced TSP

Instance	Optimal objective	Time optimal (s)	Lower bound standard	Upper bound standard	Gap standard (%)
0010	105	0	105	105	0.0
0015	271	0	271	271	0.0
0020	296	0	86	296	71.0
0025	375	0	87	386	77.5
0030	433	0	0	434	100.0
0040	458	0	0	604	100.0
0050	630	0	100	762	76.4
0060	821	0	0	948	100.0
0070	858	0	0	1.419	100.0
0080	807	0	5	1.324	100.0
0090	974	0	212	1.150	81.6
0100	996	0	266	2.228	88.1
0150	1.673	1	0	3.003	100.0
0200	2.029	0	0	3.838	100.0
0250	2.798	0	0	5.497	100.0
0300	3.695	2	119	7.240	98.4
0400	4.709	2	0	9.919	100.0
0500	5.747	5	0	9.983	100.0
0600	6.548	8	154	12.573	98.8
0700	8.097	8	0	16.169	100.0
0800	9.234	15	0	18.019	100.0
0900	9.271	12	0	19.481	100.0
1000	11.202	29	0	23.604	100.0
1500	16.339	51	0	36.304	100.0
2000	20.757	133	0	48.789	100.0
2500	1.333	510	0	35.075	100.0
3000	0	2.125	0	24.074	100.0

2 Impact on Problem Variants

We study different problem variants and evaluate the performance difference between the standard and the lifted model.

For each variant, we present a *standard MIP model.* We perform the lifting strategy as described on the balanced TSP. For the sake of simplicity, since the lifting process for each variant is totally similar to the balanced TSP, we do not present the lifted formulation alongside each standard formulation.

2.1 Balanced Assignment

A first related problem we may want to consider is the assignment problem. It consists of removing from the original problem the subtour elimination constraint. We note that it constitutes a relaxation of the original balanced TSP.

We define the balanced assignment problem as follows:

$$\begin{aligned}
&\min \quad z \\
&\text{subject to} \\
&\sum_{j \in V,\, ij \in E} x_{ij} = 2 && \forall i \in V && (11)\\
&z \geq \sum_{e \in E} x_e w_e && && (12)\\
&z \geq -\sum_{e \in E} x_e w_e && && (13)\\
&x_e \in \{0,1\} && \forall e \in E \\
&z \in \mathbb{Z}
\end{aligned}$$

Constraint (11) forces each vertex to have two incident edges, constraint (12) and (13) force the balanced objective.

As for the balanced TSP, one can define the lifted version of the balanced assignment.

2.2 Balanced Fixed-Size Subset

In this section, we describe a relaxation of the assignment problem. We relax the degree constraint on each vertex and add a constraint to fix the number of edges selected. The following model describes the subset problem:

$$\begin{aligned}
&\min \quad z \\
&\text{subject to} \\
&\sum_{e \in E} x_e = n && && (14)\\
&z \geq \sum_{e \in E} x_e w_e && && (15)\\
&z \geq -\sum_{e \in E} x_e w_e && && (16)\\
&x_e \in \{0,1\} && \forall e \in E \\
&z \in \mathbb{Z}
\end{aligned}$$

Constraint (14) forces the number of edges selected, constraints (15), (16) define z as an objective. As for the previous models, we consider both the standard version and the lifted version.

2.3 TSP

Since the lifted formulation for the balanced-(TSP, assignment, subset) made a huge difference in the performance, we added some experiments for the classical TSP. Without surprise, gurobi was able to handle even the big instances in a few seconds for both models. However, it seems that the lift formulation does not imply any gain in performance on the classical TSP. The following model presents the TSP formulation used:

$$\min \quad z$$
$$\text{subject to}$$
$$\sum_{j \in V, ij \in E} x_{ij} = 2 \qquad \forall i \in V \tag{17}$$
$$\sum_{i,j \in S, ij \in E} x_{ij} \leq |S| - 1 \qquad \forall S \subset V, S \neq \emptyset \tag{18}$$
$$z \geq \sum_{e \in E} x_e w_e \tag{19}$$
$$x_e \in \{0,1\} \qquad \forall e \in E$$
$$z \in \mathbb{Z}$$

Constraints (17) and (18) are tour constraints. Constraint (19) sets the objective to the minimal tour value.

2.4 Balanced Spanning Tree

A related problem can be to find a balanced minimum spanning tree in the graph. We relax the degree constraint for each vertex. Also, we add a constraint that forces the number of selected edges to be $n-1$ and the graph to be connected. The following model describes the balanced spanning tree problem:

$$\min \quad z$$
$$\text{subject to}$$
$$\sum_{e \in E} x_e = n - 1 \tag{20}$$
$$\sum_{i,j \in S, ij \in E} x_{ij} \leq |S| - 1 \qquad \forall S \subset V, S \neq \emptyset \tag{21}$$
$$z \geq \sum_{e \in E} x_e w_e \tag{22}$$
$$z \geq -\sum_{e \in E} x_e w_e \tag{23}$$
$$x_e \in \{0,1\} \qquad \forall e \in E$$
$$z \in \mathbb{Z}$$

Constraint (20) ensures the number of selected edges to be $n-1$. Constraints (21) ensures no cycle in the selected edges. Constraints (22) and (23) forces the objective function to be the balanced sum of selected edges.

2.5 Numerical Results on Sub-cases

As for the balanced TSP, results were obtained on an Intel(R) Core(TM) i5-3470 CPU @ 3.20GHz core with 8 GB RAM and Linux Mint 11. The MIP models were implemented using Gurobi 8.1 and python2.7. Tables 2, 3, 4 and 5 present numerical results for respectively the Balanced assignment problem, the balanced spanning tree problem, the balanced subset problem and the TSP.

For the balanced assignment problem and the balanced subset problem, the result tables are similar to the balanced TSP (*i.e.* time to optimal with the lifted formulation and the bounds and gap for the standard formulation). For the classical TSP, since both formulations were able to prove optimality within a few seconds, we only use the time to optimal for both formulations. Finally, for the balanced spanning tree, since both formulations struggle to find optimality proofs for some instances, we stopped them if they run more than 30 s. We report the bounds of both formulations.

We note that for both the balanced assignment and the balanced subset, we obtain similar results as for the balanced TSP. For the classical TSP, we do not notice any significant difference between the two formulations. More tests should be performed on different non-balanced problems to evaluate the efficiency (or inefficiency) of the lifted formulation. Finally, the balanced minimum spanning tree is harder to solve than the other versions (including the balanced TSP). Indeed, even for some very small instances ($n = 40$), the lifted formulation is not able to prove optimality within 30 s. We note that the biggest solved instance within 30 s with the lifted formulation has 300 vertices. Also, the standard formulation is only able to solve the smallest instance ($n = 10$). Even for $n = 15$, the standard formulation is not able to find an optimal lower bound nor an optimal upper bound.

3 Conclusion and Perspectives

This paper presents a new MIP model for the balanced TSP. This model uses a specific property on the edge weights. This allows to make a clever reasoning and closing instances of the Metaheuristic Summer School (MESS18) competition. This paper investigates this phenomenon on several variants of the balanced TSP (namely the balanced assignment problem, the balanced fixed subset problem, the balanced minimum spanning tree problem and the classical TSP). We show that this performance improvement occurs on all balanced problems considered. However, this phenomenon seems to not appear on the classical TSP. We showed a dramatic performance increase on the lifted formulation compared to the standard formulation. This approach seems to be suited for balanced problems and a simple MIP model is even able to compete with other metaheuristics presented during the competition.

MIPs are known to be suited for instances of reasonable size. However, this approach is less suited on bigger instances (*i.e.* 5.000, 10.000, *etc.*). One way to overcome this issue is to implement *matheuristics*. One can, for instance, implement some Large Neighbourhood Search while using a MIP solver [FL03]. This, way, one can take advantage of the progress made by such software and the lifted formulation described in the present paper.

A Appendix

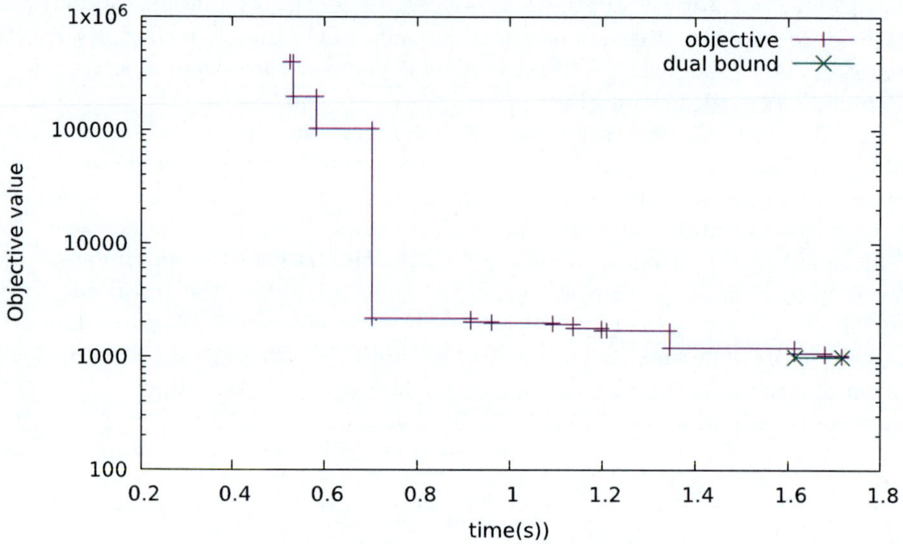

Fig. 1. btsp 0100 convergence curves

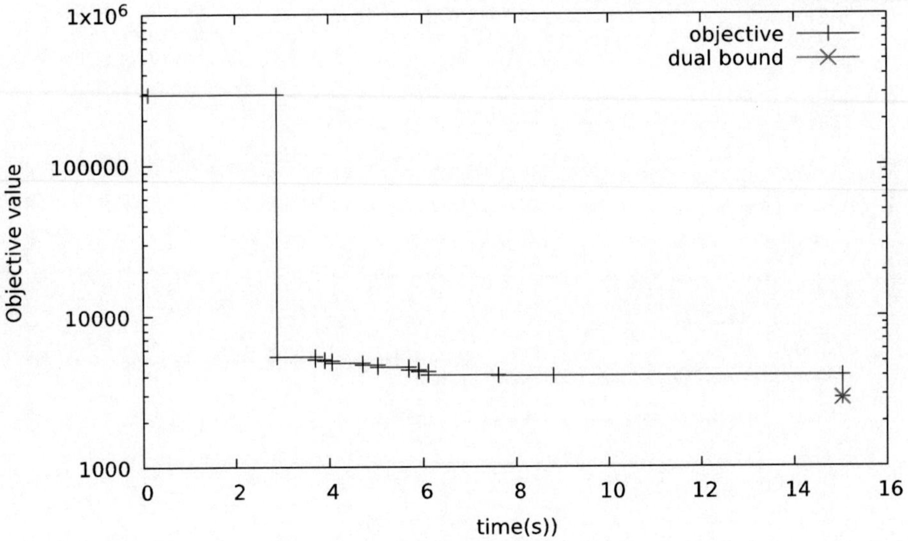

Fig. 2. btsp 0250 convergence curves

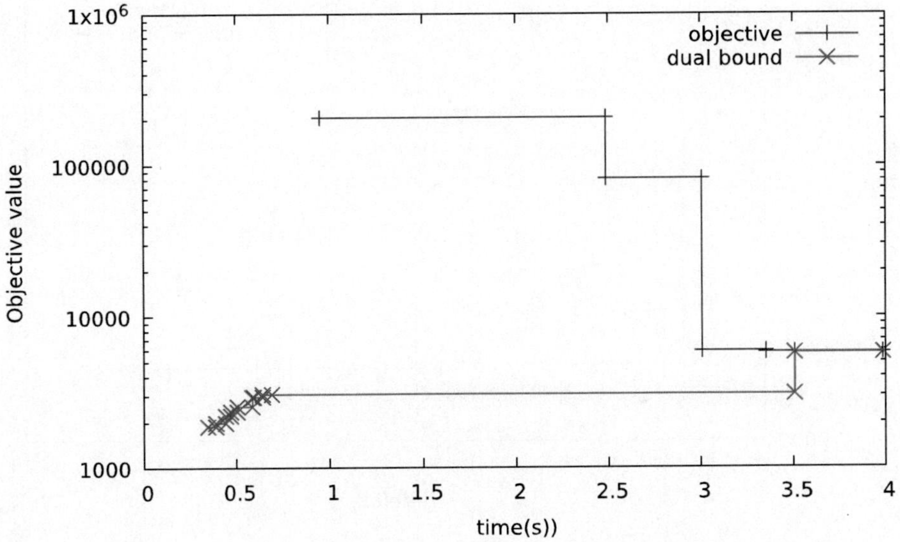

Fig. 3. btsp 0500 convergence curves

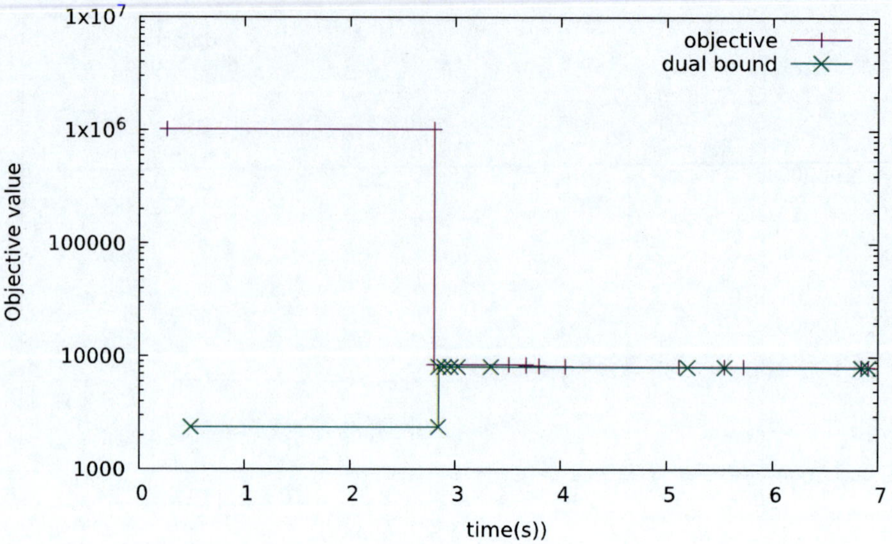

Fig. 4. btsp 0700 convergence curves

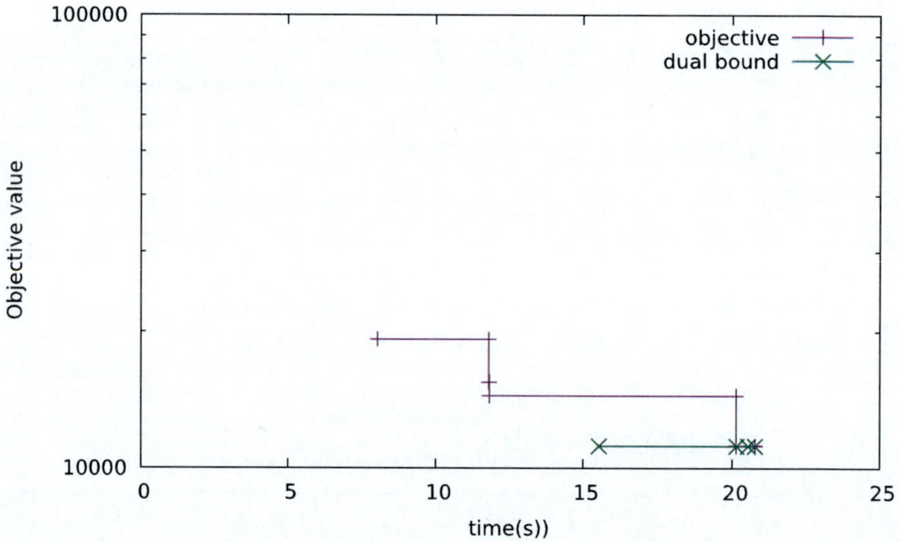

Fig. 5. btsp 1000 convergence curves

Table 2. Numerical results on the balanced assignment problem

Instance	Time to optimal (s)	Optimal objective	Lower bound standard	Upper bound standard	Gap standard (%)
0010	0	105	105	105	0.0
0015	0	271	271	271	0.0
0020	0	285	221	286	22.7
0025	0	364	214	375	42.9
0030	0	433	161	443	63.6
0040	0	458	56	457	87.7
0050	0	629	117	805	85.4
0060	0	821	50	1.083	95.3
0070	0	857	90	1.210	92.5
0080	0	806	225	850	73.5
0090	0	973	58	1.203	95.1
0100	0	993	8	1.399	100.0
0150	0	1.664	0	2.868	100.0
0200	0	2.029	1	3.936	100.0
0250	0	2.796	0	5.056	100.0
0300	0	3.693	141	6.721	97.5
0400	1	4.698	0	7.099	100.0
0500	1	5.737	0	8.922	100.0
0600	0	6.543	4	9.789	99.9
0700	1	8.095	0	13.028	100.0
0800	1	9.226	0	14.899	100.0
0900	2	9.265	0	15.412	100.0
1000	2	11.201	0	21.157	100.0
1500	16	16.337	0	31.883	100.0
2000	18	20.755	0	47.645	100.0
2500	39	1.326	0	31.943	100.0
3000	38	0	0	18.872	100.0

Table 3. Numerical results on the balanced spanning-tree problem

Instance	Lited lower bound	Lifted upper bound	Standard lower bound	Standard upper bound
0010	60	60	60	60
0015	129	129	4	173
0020	143	143	0	242
0025	222	222	0	430
0030	247	247	0	247
0040	241	840	0	720
0050	378	378	0	1.036
0060	389	394	0	1.071
0070	375	375	0	1.238
0080	369	369	0	1.167
0090	342	342	0	1.300
0100	525	525	0	2.313
0150	855	855	0	2.476
0200	925	925	0	4.311
0250	1.170	1.170	0	3.236
0300	1.471	1.471	0	4.854
0400	2.099	2.109	0	7.493
0500	2.619	2.629	0	6.942
0600	0	2.917	0	10.381
0700	0	3.864	0	15.749
0800	0	4.319	0	15.649
0900	325	4.189	0	18.736
1000	386	5.188	0	21.090
1500	0	20.396	0	32.876
2000	0	23.264	0	45.489
2500	0	24.583	0	30.485
3000	0	28.297	0	13.466

Table 4. Numerical results on the balanced subset problem

Instance	Time to optimal (s)	Optimal objective	Lower bound standard	Upper bound standard	Gap standard (%)
0010	0	82	0	82	100.0
0015	0	130	0	130	100.0
0020	0	144	92	332	72.2
0025	0	233	0	306	100.0
0030	0	259	0	434	100.0
0040	0	263	0	460	100.0
0050	0	389	0	700	100.0
0060	0	406	128	568	77.4
0070	0	376	0	650	100.0
0080	0	358	0	1.750	100.0
0090	0	353	0	486	100.0
0100	0	537	0	2.074	100.0
0150	0	799	0	3.519	100.0
0200	0	926	0	4.176	100.0
0250	0	1.192	0	5.032	100.0
0300	0	1.493	0	5.785	100.0
0400	0	2.110	0	8.620	100.0
0500	0	2.630	0	10.708	100.0
0600	0	2.861	0	12.612	100.0
0700	1	3.886	0	14.463	100.0
0800	0	4.320	0	15.765	100.0
0900	0	4.166	0	16.042	100.0
1000	9	5.210	0	21.334	100.0
1500	19	7.579	0	33.974	100.0
2000	34	8.447	0	41.608	100.0
2500	7	0	0	29.713	100.0
3000	0	0	0	16.087	100.0

Table 5. Numerical results on the classical TSP problem

Instance	Time to optimal (s)	Standard time to optimal (s)
0010	0	0
0015	0	0
0020	0	0
0025	0	0
0030	0	0
0040	0	0
0050	0	0
0060	0	0
0070	0	0
0080	0	0
0090	0	0
0100	0	0
0150	0	0
0200	0	0
0250	0	0
0300	0	0
0400	0	0
0500	0	0
0600	0	0
0700	0	0
0800	0	0
0900	0	0
1000	1	1
1500	0	0
2000	2	2
2500	4	3
3000	2	2

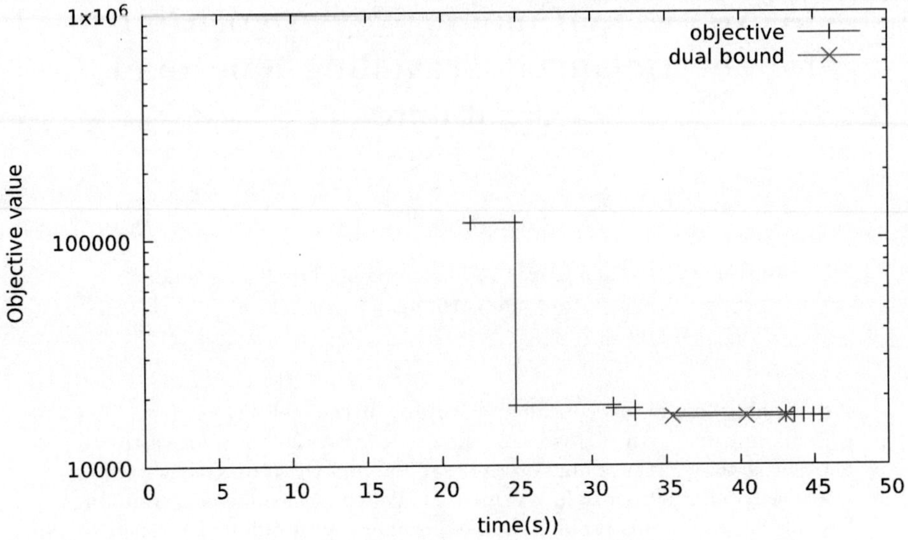

Fig. 6. btsp 1500 convergence curves

Reference

FL03. Fischetti, M., Lodi, A.: Local branching. Math. Program. **98**(1–3), 23–47 (2003)

A Memetic Random Key Algorithm for the Balanced Travelling Salesman Problem

Ayse Aslan(✉)

University of Groningen, 9747 Groningen, The Netherlands
ayse.aslan@rug.nl

Abstract. This paper considers a variant of the well-known travelling salesman problem. In this variant, the cost of travelling from a vertex to another is an arbitrary value on the real line and the objective is finding a tour with minimum absolute value cost. We propose a memetic random key algorithm for this problem and experiment with different settings of the algorithm. The experiments explore the use of a flexible decoding mechanism and also the frequency of applying local search within the random key algorithm.

Keywords: Memetic algorithms · Local search · Travelling salesman problem

1 Introduction

The travelling salesman problem is one of the most exhaustively studied combinatorial optimization problems. This problem aims to find a shortest-length tour of a given set of cities that visits each city exactly once. This tour is sometimes called a *Hamiltonian* cycle. It is proven that the travelling salesman problem is NP-hard. For instance, consider that we represent a solution of this problem by the order that cities are to be visited, then with this representation for a problem with k many cities, the solution space consists of $k!$ many solutions, if it is possible to visit any two cities consecutively. That's why it is very challenging to find optimal solutions of this problem with large number of cities as solution space grows at least exponentially in the size of the instance, the number of cities. The first important achievement made in tackling the large instances of this problem is made in 1954 by Dantzig et al. [2], in solving an instance with 49 cities. The review by Laporte [3] gives a great overview of the both exact and heuristic methods proposed and tested for this intractable problem.

In this paper, we present a variant of the travelling salesman problem, which we name *the balanced travelling salesman problem*. Now, let's formally define our problem. Let $G = (V, E)$ be a graph of n many vertices V and of their connecting edges E. Consider that a cost c_{ij} is defined for any vertex $i \in V$ and $j \in V$ such that there is an edge $e \in E$ that connects i and j in the graph G. In the travelling

S. Greco et al. (Eds.): MESS 2018, AISC 1332, pp. 16–22, 2021.
https://doi.org/10.1007/978-3-030-68520-1_2

salesman problem, the costs $c_{i,j}$ are defined to be non-negative values, as often they represent the distances between two locations. However, in the balanced travelling salesman problem, the costs are arbitrary values on the real line $\mathbb{R}$; they can take also negative values. The objective in the usual travelling salesman problem is to find a Hamiltonian cycle with minimum tour length. However, in the problem discussed here, the objective is to find a Hamiltonian cycle such that the absolute value of its tour length is minimum. For convenience, we also name the absolute length of a tour, the "cost" of the tour, in this document.

For this problem, we propose a random key memetic algorithm. Random key algorithms are already proposed for some travelling salesman problems in the literature [4] and memetic algorithms also [1].

2 The Memetic Random Key Algorithm

The random keys π_i are defined for each vertex $i \in V$ that takes a continuous value in the interval of $[0, 1]$. These values represent the visit order priorities of vertices. For instance, if $\pi_i > \pi_j$ for some vertices i and j, then we can interpret this as vertex i has the priority to be visited before vertex j.

Algorithm 1. Pseudocode of the Memetic Random Key Algorithm

$P \leftarrow \emptyset$;
Initialize $n_{top}, n_{cx}, n_{mig}, p_{elit} \in [0,1], p_{ls} \in [0,1]$;
$P \leftarrow GenerateInitialSolutions(n)$;
$P \leftarrow ApplyLocalSearch(P, p_{ls})$;
while time limit not reached **do**
  $P_{top} \leftarrow SelectTopSolutions(P, n_{top}), P_{cx} \leftarrow \emptyset, P_{mig} \leftarrow \emptyset$;
  for $m \in \{1, .., n_{cx}\}$ **do**
    select randomly s_a and $s_b \in P$ that $s_a \neq s_b$;
    $s_{new} \leftarrow BiasedUniformCrossover(p_{elit}, s_a, s_b)$;
    $s_{new} \leftarrow ApplyLocalSearch(s_{new}, p_{ls})$;
    $P_{cx} \leftarrow P_{cx} \cup \{s_{new}\}$;
  end for
  $P_{mig} \leftarrow GenerateInitialSolutions(n_{mig})$;
  $P \leftarrow P_{top} \cup P_{cx} \cup P_{mig}$;
end while
return P

The pseudocode of the memetic algorithm is given in Algorithm 1.

Initial Population: The solutions of initial population are generated by constructing n tours ($GenerateInitialSolutions(n)$). Each tour is constructed in the following way. Firstly, a vertex is selected randomly to be the starting vertex of the tour. Then, at each iteration a new vertex is added to the tour, a vertex that is not added to the tour yet. The new vertex to add to the tour is randomly selected amongst the neighbors of the most recently added vertex. This tour construction process is repeated until there are no more vertices left that can be added to the tour. Lastly, the remaining vertices are added to the tour in a random fashion.

Local Search: Two local search methods are utilized in the memetic algorithm. Similarly to [4], we also use one 2-opt and one swap operator based local search methods that are not integrated; specifically, here we apply first the swap, then the 2-opt local search on a given solution. These methods are not fixed to be used on every generated solution. They are applied each time with probability p_{ls}. The effect of this parameter is investigated in the experiments. The pseudocode of the $ApplyLocalSearch()$ procedure is described in Algorithm 2.

Algorithm 2. Pseudocode of the $ApplyLocalSearch()$ procedure

$s_a, p_{ls}, Niter$ are given;
toss a coin c that takes value in $[0, 1]$;
if $c \leq p_{ls}$ **then**
    for $m \in \{1, ..., Niter\}$ **do**
        for $i \in V$ and $j \in V, j > i$ **do**
            swap i and j in the tour of solution s_a, if this reduces the cost;
        end for
    end for
    for $m \in \{1, ..., Niter\}$ **do**
        for $i \in V$ and $j \in V, j > i + 1$ **do**
            apply 2-opt operator on ith and jth places of the tour of solution s_a, if this reduces the cost;
        end for
    end for
end if

Population Update: Each time a population is renewed, a number of (n_{top}) solutions in the current population which have the lowest cost values are directly transferred to the new population, a number of (n_{cx}) solutions are generated by the crossover operator applied on the solutions in the current population and finally a number of (n_{mig}) solutions are newly generated and added to the new population. These solutions are generated by $GenerateInitialSolutions()$ procedure.

Crossover: The biased uniform crossover is used in the memetic algorithm. This crossover randomly selects two parent solutions in the population to produce a single offspring solution. The produced offspring takes genes of the parents one-by-one. For each gene $\pi_i, i \in V$, a biased coin (p_{elit}) is tossed to select the parent of the gene to be transmitted to the offspring. The parameter p_{elit} represents the bias of favoring the parent solution with lower cost.

Decoders: The algorithm proposed here uses an indirect representation, the priorities of vertices. This paper suggests two different decoders where any of them can be used to decode a given random key solution. The second one is a flexible version of the first one.

- Decoder 1: For the given random key solution $\pi_1, ..., \pi_k$, sort the random keys in descending order. The decoded tour solution is the vertices in the order of descending random key values. For example, consider an instance with five vertices; 1, 2, 3, 4 and 5. Let's say that the random key solution is $0.1, 0.2, 0.45, 0.15, 0.25$, then the decoded tour is $3 \rightarrow 5 \rightarrow 2 \rightarrow 4 \rightarrow 1$.
- Decoder 2: This decoder is more flexible than the previous. The first decoder does not consider the feasibility of the suggested tour by the priority levels. For instance, in the above example if let's say vertex 5 is not a neighbor of vertex 3 and however if vertex 2 is a neighbor of both vertex 3 and 5, then the tour $3 \rightarrow 2 \rightarrow 5 \rightarrow 4 \rightarrow 1$ could be a better solution than the tour decoded by the first decoder. This decoder, using this idea, applies the same decoding rule given in the first decoder until in the constructed tour it is feasible to add vertices with respect to their neighbor relations, edges in the graph G, and after that it adds the first vertex feasible in the descending order of vertices, with respect to their random keys. The purpose of proposing this flexible decoder is to investigate if this type of decoding is useful in being able to obtain feasible solutions.

3 Computational Experiments

In the experiments, the instances of the metaheuristic competition declared in MESS 2018 Summer School (https://www.ants-lab.it/mess2018/) are used. These instances are varied in the number of vertices, from 10 to 3000 vertices. The graphs of these instances are not fully connected; the number of edges in each instance is four times of the number of vertices, $|E| = 4|V|$.

The memetic random key algorithm is implemented in C++ language and tested on these instances under a time limit of five hours. The algorithm is offline tuned to have $N_{iter} = 100$ for instances with a number of vertices less than or equal to 100 and the rest of the instances are tuned to implement 1000 of local search iterations. Also the population composition of the algorithm is tuned offline; in all instances n_{top}, n_{cx} and n_{mig} are tuned to 20, 120 and 60, respectively. Namely, the population size is 200 in all instances. The elitist probability p_{elit} is also among the parameters that are offline tuned; this is tuned to 0.6 in all instances.

The experiments conducted here focus on two parameters: the frequency of applying local search (p_{ls}) and the version of decoder to use each time a new solution is produced in the memetic random key algorithm. For each of these parameters, three settings are experimented on the instances to understand their effect on the performance of the memetic algorithm.

For each parameter setting the memetic algorithm is run for 10 times. The results by the algorithm show that the algorithm is not able to find feasible solutions for the large instances, the instances with $|V| > 100$. Therefore, only the results on the remaining instances are to be showed in this section.

One-way ANOVA is used in all experiments for detecting significances between means of different settings.

3.1 The Decoder Effect

The aim is to understand if having a flexible decoder, Decoder 2, can be useful sometimes or not. For this purpose, three different decoding settings are tested. The first setting applies Decoder 2 always, the second setting applies Decoder 2 half the time, with the help of the unbiased coin tossing, and applies Decoder 1 otherwise. Lastly, the third setting does not use the flexible decoder at all; it applies Decoder 1 always on every generated solution. In the experiments of these settings, the local search frequency parameter p_{ls} is fixed to 0.3.

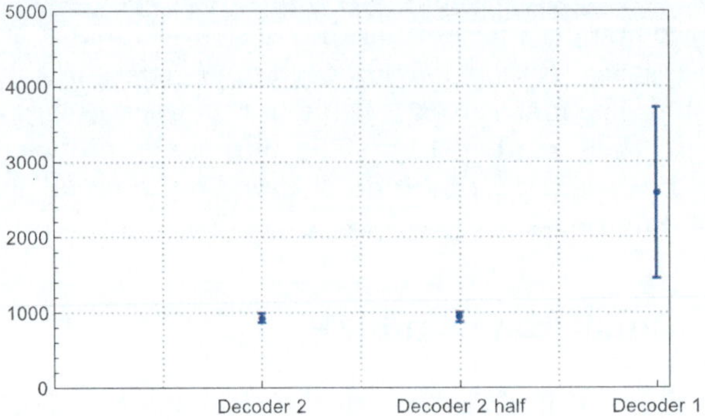

Fig. 1. The means plot of decoder settings with 95% confidence intervals

The Fig. 1 shows the cost means of the tested three decoder parameter settings with error bounds. The means seem better for the settings that make use of the flexible decoder, Decoder 2. However, statistically, the difference between means are not found important with the confidence level of 95% (p-value = 0.08). This can be due to the low number of samples.

3.2 The Local Search Frequency Effect

The parameter p_{ls} determines how frequently produced solutions are to be applied local search in the memetic algorithm. Three levels of this parameter are tested; 0.1, 0.3 and 0.5. In this experiment, the decoder method is fixed to the second setting experimented in Sect. 3.1, the method that applies Decoder 1 or Decoder 2 with equal probability each time.

The Fig. 2 shows the mean costs of the tested three settings with 95% confidence error bounds. Again, the effect of this parameter is not found statistically significant (p-value = 0.37) with 95% confidence level, probably due to the low number of samples taken of the settings. However, the means may indicate that higher local search frequencies are preferred in the memetic algorithm.

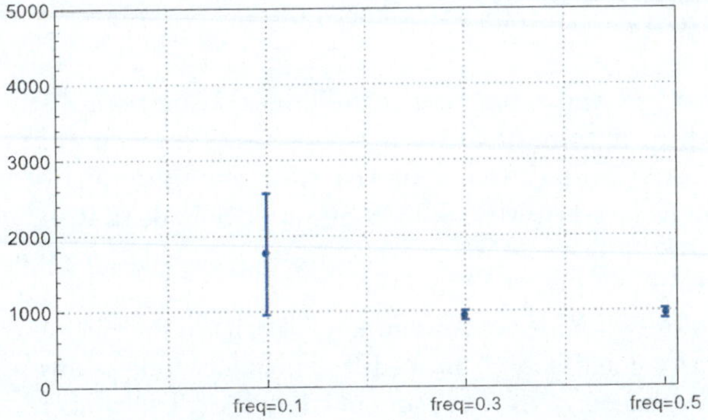

Fig. 2. The means plot of local search frequency settings with 95% confidence intervals

3.3 Solutions

The solutions found of the instances are presented in this section for bechmark purposes. These are given in Table 1. The first values in the cells of the table give the best found solutions, while the second values give the averages of the solutions found in the runs. The decoder settings are defined by labels *dec_set* and the local search frequencies are defined by *freq_set* in the table. These labels are given numbers of 1, 2 or 3, which define the specific setting order considered in Sect. 3.1 and Sect. 3.2. The instances are defined by the number of vertices in them.

Table 1. Best and average solutions by the different settings of the memetic algorithm

Instances	*dec_set* = 1, *freq_set* = 2	*dec_set* = 2, *freq_set* = 1	*dec_set* = 2, *freq_set* = 2	*dec_set* = 2, *freq_set* = 3	*dec_set* = 3, *freq_set* = 2
10	105, 105	105, 105	105, 105	105, 105	105, 105
15	271, 271	271, 271	271, 271	271, 271	271, 271
20	296, 296	296, 296.1	296, 296	296, 296	296, 296
25	376, 376.1	375, 375	376, 376.2	375, 376.4	375, 375
30	499, 506.6	510, 511.2	467, 478.1	490, 497.6	500, 505.4
40	614, 614	627, 627	615, 629.1	591, 591	658, 664.3
50	850, 890.9	915, 941	902, 926.1	883, 897.2	927, 946.8
60	1130, 1150.9	1176, 1178.8	1162, 1167.7	1172, 1182.4	1209, 1211.7
70	1308, 1341.3	1377, 1402.5	1365, 1365	1396, 1417.4	1378, 1378
80	1542, 1584.3	1568, 1654.6	1545, 1590.8	1565, 1609.3	1764, 1816.1
90	1746, 1889.5	1860, 1863.2	1834, 1876.6	1881, 1950.9	2125, 2160.1
100	2136, 2203.2	2274, 11892.8	2177, 2232.5	2037, 2345.3	2316, 21350.5

4 Conclusions

This paper considers the balancing travelling salesman problem in which the objective is to find a tour of a given set of vertices such that each vertex is exactly once visited and the cost of the tour is minimum in absolute value. This problem is declared in MESS 2018 Summer School that is held in Acireale, Sicily in July, 2018 for the metaheuristic competition of the school. The instances of this competition are used in the experiments of this paper. In this paper, we propose and explore a memetic random key algorithm for this new problem. This algorithm uses two local search procedures in which one is swap and the other is 2-opt operator based. The experiments focus on the decoder function used for interpreting the indirect random key solutions that take continuous values in $[0, 1]$ and on the frequency of the local search applied on the newly produced solutions by the algorithm. The experiment results show some indications of favoring flexible decoding mechanisms. With respect to local search frequency, the moderate levels ($p_{ls} \in \{0.3, 0.5\}$) seem to be more suitable than the low level ($p_{ls} = 0.1$).

References

1. Bontouxa, B., Artigues, C., Feillet, D.: A memetic algorithm with a large neighborhood crossover operator for the generalized traveling salesman problem. Comput. Oper. Res. **37**, 1844–1852 (2010)
2. Dantzig, G., Fulkerson, R., Johnson, S.: Solution of a large-scale travelling-salesman problem. J. Oper. Res. Soc. Am. **2**(4), 393–410 (1954). https://doi.org/10.1287/opre.2.4.393
3. Laporte, G.: The traveling salesman problem: an overview of exact and approximate algorithms. Eur. J. Oper. Res. **59**, 231–247 (1992)
4. Snyder, L.V., Daskin, M.S.: A random-key genetic algorithm for the generalized traveling salesman problem. Eur. J. Oper. Res. **174**, 38–53 (2006)

A Variable Neighborhood Search Algorithm for Cost-Balanced Travelling Salesman Problem

Mehmet A. Akbay and Can B. Kalayci(✉)

Department of Industrial Engineering, Pamukkale University, 20160 Denizli, Turkey
cbkalayci@pau.edu.tr

Abstract. Travelling salesman problem (TSP) can be described as finding minimum Hamilton cycle cost for a network consisting of starting/ending nodes and intermediate nodes. It is one of the most practiced classical combinatorial optimization problems due to its several application areas as well as its convertibility to various problem types. For this reason, many researchers and practitioners have studied on several variations of the problem. Despite its simple structure, obtaining an exact solution becomes harder as the problem dimension increases. Therefore, heuristic algorithms have been widely adopted by various researchers in order to obtain near optimal solutions. Within the scope of this study, one of the recent variants of TSP which is known as Cost-Balanced TSP is considered and a solution approach based on the variable neighborhood search algorithm has been proposed. Computational experiments have been performed on twenty-two publicly available datasets which includes small, medium and large-scale instances. Efficiency of proposed solution approach has been discussed according to the computational results.

Keywords: Travelling salesman problem · Cost-balanced · Negative edge cost · Metaheuristics · Variable neighborhood search

1 Introduction

Travelling salesman problem (TSP) formulated by Menger (1932) has become one of the most studied combinatorial optimization problems since it was firstly introduced to the literature. It can be simply described as finding the optimal route beginning from a single starting node and ending at the same node by visiting each node only once. In other words, TSP searches for the minimum Hamilton cycle cost for a network consisting of starting/ending and intermediate nodes. The distance between two nodes is the same in each opposite direction, forming an undirected graph.

Let $G = (N, A)$ with a node set $N = \{0, 1, 2, \ldots, n\}$ and an edge/arc set $A = \{(i, j), i, j \in N\}$. And let $C = \left(c_{ij}\right)$ be the distance or cost matrix according to the problem type.

Within this scope, classical TSP has two main objectives;

- finding an appropriate node subset $S \subseteq N$,

S. Greco et al. (Eds.): MESS 2018, AISC 1332, pp. 23–36, 2021.
https://doi.org/10.1007/978-3-030-68520-1_3

- finding a minimum cost Hamiltonian cycle.

Over the years, several researchers have studied numerous variants (Table 1) of TSP including feasible and infeasible Hamiltonian Cycles (Fischetti et al. 2007).

Since TSP is the core of many real-life optimization problems, it has drawn a great attention of researchers and practitioners on a wide range of application areas such as planning, scheduling, logistics, manufacturing, etc. Some of these application areas are listed as follows: drilling of printed circuit boards (Grötschel et al. 1991), X-Ray crystallography (Bland and Shallcross 1987), overhauling gas turbine engines (Plante et al. 1987), the order-picking problem in warehouses (Ratliff and Rosenthal 1983), controlling of robot motions (Jünger et al. 1995), computer wiring, mask plotting in PCB production, vehicle routing (Lenstra and Kan 1975). Many problem types can be easily described as one of the TSP variants.

Since it is an NP-complete problem (Karp 1972), it may not possible to obtain an optimal solution for all instances of the problem within a polynomial time. While exact solution methods have been widely used to solve relatively smaller instances and/or to test new models (Kinable et al. 2017; Applegate et al. 2003), heuristic algorithms have been adopted by several researchers to get near optimal solutions since it is getting difficult to obtain an optimal solution as problem dimension increases (Laporte 1992).

By this means, recent studies on solving TSP by heuristic solution approaches are briefly reviewed and presented in Table 2.

Escario et al. (2015) has redesigned the ant colony optimization algorithm, originally proposed by Dorigo and Gambardella (1997) for solving TSP, based on population dynamics mechanism which allows to decide number of ants to be recruited for the further steps according to the previous ant performance. Ismkhan (2017) used pheromone information in local search phase to overcome deficiencies such as high time and space-complexity of ACO on solving TSP.

Wang et al. (2016) proposed a multi-offspring genetic algorithm, originally developed by Holland (1975), for enabling faster calculation speed as well as a smaller number of iterations to achieve better solutions of TSP compared to the original genetic algorithm scheme. Hore et al. (2018) improved classical VNS algorithm with a stochastic approach in order to escape from local optimum. Meng et al. (2017) developed a variable neighborhood search algorithm with two-stage greedy initialization. In order to avoid duplicated routes, they adopted direct route encoding mechanism to the algorithm to solve colored TSP.

Many researchers have developed several heuristic and metaheuristic approaches involving efficient exploration and exploitation features with the aim of achieving improved algorithmic performance to solve various TSP variants. In this paper, to solve cost-balanced TSP, and a candidate solution approach based on a variable neighborhood search (VNS) algorithm has been proposed.

The remainder of this paper is organized as follows: Sect. 2 presents the definition of the Cost-Balanced TSP including basic assumptions and mathematical formulation. While Sect. 3 presents the proposed methodology, Sect. 4 presents the computational experiments and results. Finally, Sect. 5 concludes the paper with a short discussion and future research directions.

Table 1. TSP variants and publications

Problem type	Reference	Year	Definition
Asymmetric TSP	Carpaneto and Toth (1980)	1980	Paths may not exist in both directions and the distances might be different, forming a directed graph
The Traveling Purchaser Problem	Ramesh (1981)	1981	Considering a network consisting of some markets selling different goods with different prices, objective is to find the optimal tour for minimum purchasing and travel cost
The Prize-Collecting TSP	Balas and Martin (1985)	1985	Each node has an associated prize, and cost. Objective is to obtain the cycle with nodes having minimum cost and total price is not less than given threshold value
The Orienteering Problem (Selective TSP)	Laporte and Martello (1990)	1990	Objective is to create a route/cycle maximizing total profit. Total length/cost cannot exceed predefined threshold
Maximum collection problem	Butt and Cavalier (1994)	1994	Maximizing total rewards collected at the nodes visited
The Covering Tour Problem	Gendreau et al. (1997)	1997	Finding a minimum length/cost Hamiltonian cycle where all nodes of network have been covered by the tour
The Generalized TSP	Fischetti et al. (2007)	1997	Nodes are separated into clusters. Objective is to find minimum cost Hamiltonian cycle for each cluster
The Median Cycle Problem	Labbé et al. (1999)	1999	Finding an optimal cycle visiting a given vertex considering routing and assigning cost
Multiple TSP	Bektas (2006)	2006	A generalization of the TSP in which more than one salesman is allowed

(*continued*)

Table 1. (*continued*)

Problem type	Reference	Year	Definition
The Capacitated Prize-Collecting TSP	Tang and Wang (2008)	2008	Each node has a prize, a penalty and a demand, and can be visited only once. The objective is to minimize the total distance travelled or cost and the net penalties and maximize prize collected. The demand of visited nodes cannot exceed the salesman's capacity

2 Cost-Balanced TSP

Cost balanced TSP is one of the recent variants of the TSP. Basic assumptions of the problem can be stated as shown below:

- The main objective is to find a Hamiltonian cycle of which total travel cost as close as to 0.
- Like classical TSP, cost balanced TSP has a symmetric cost/length matrix.
- Negative edge cost is allowed.

Relevant problem can be mathematically modelled by simply revising the original objective of the problem to its modulus.

An integer linear programming formulation for the cost-balanced TSP is given with the following notation by extending the classical TSP formulation based on (Dantzig et al. 1954) which is one of the most cited mathematical model in the TSP literature.

Indices:

$i, j \in N$ the set of nodes

Parameters:

c_{ij} cost of related edge

Minimize

$$\left| \sum_{i<j} c_{ij} x_{ij} \right| \tag{1}$$

Subject to

$$\sum_{i<k} x_{ik} + \sum_{j>k} x_{kj} = 2 \qquad k = 1, \ldots, n \tag{2}$$

$$\sum_{i,j \in S} x_{ij} \leq |S| - 1 \qquad S \subset \mathrm{N}, 3 \leq |S| \leq n - 3 \tag{3}$$

$$x_{ij} \in \{0, 1\} \tag{4}$$

$$i, j = 1, \ldots, n, i < j.$$

While objective function (1) aims to minimize the absolute value of total cost, equality (2) refers degree constraints and inequality (3) refers sub-tour elimination constraints. Finally, constraint (4) imposes that x_{ij} is binary.

Table 2. Recent studies using heuristic methods for some of the TSP variants

Problem type	Solution approach	Originally proposed	Publications
TSP, Dynamic TSP	Ant Colony Optimization	Dorigo et al. (1996)	(Dahan et al. 2019; Mavrovouniotis et al. 2017; Escario et al. 2015; Dorigo and Birattari 2010; Ismkhan 2017)
TSP	Artificial Bee Colony Optimization	Karaboga (2005)	(Shokouhifar et al. 2015; Kocer and Akca 2014; Zhong et al. 2017)
TSP, Minmax Multiple TSP, Symmetric/Asymmetric TSP, Orienteering problem, Colored TSP, bi-objective TSP	Variable Neighborhood Search	Mladenovic and Hansen (1997)	(Palomo-Martínez and Salazar-Aguilar 2019; Meng et al. 2017; Hore et al. 2018; Lahyani et al. 2017; Sevkli and Sevilgen 2006; Wang et al. 2017; Mjirda et al. 2017)
TSP	Particle Swarm Optimization	Kennedy and Eberhart (1995)	(Akhand et al. 2014)
The Capacitated Prize-Collecting TSP	Iterated Local Search	Lourenço et al. (2003)	(Tang and Wang 2008)
TSP	Cuckoo Search	Gandomi et al. (2013)	(Ouaarab et al. 2014)
Asymmetric TSP	Harmony Search	Geem et al. (2001)	(Boryczka and Szwarc 2019b, a)
Symmetric/Asymmetric TSP	Bat Algorithm	Yang (2010)	(Saji and Riffi 2016; Osaba et al. 2016)
Symmetric TSP, Orienteering problem	Evolutionary Algorithms	Holland (1975)	(Kobeaga et al. 2018; Wang et al. 2016)
TSP	Invasive Weed Optimization	Mehrabian and Lucas (2006)	(Zhou et al. 2015)

3 Proposed Methodology: A Variable Neighborhood Search Algorithm

Large-scale instances of TSP cannot be solved within polynomial time with exact solution approaches. Hence, in order to solve cost-balanced TSP, a VNS based solution approach has been proposed.

VNS is one of the recent metaheuristics which was developed for combinatorial and global optimization problems (Mladenovic and Hansen 1997). To acquire better searching performance, VNS systematically applies multiple neighborhood structures (Hansen and Mladenovic 2001). Using multiple neighborhood structure is one of the most significant and important features of VNS which distinguishes it from other local search-based heuristics.

Former studies have shown that local search-based algorithms can often be stuck in local optimum points, whereas shaking step of VNS algorithm diversifies the search systematically with the aim of preventing the algorithm to stuck into local optima.

General procedure of the VNS algorithm is given in Fig. 1. The VNS algorithm is basically composed of two main parts: shaking and local search. In the first step of the algorithm, the neighborhood structures to be used in the shaking and local search steps are determined. At this stage, it is of a great importance to identify neighborhood structures that can effectively explore search space. Since the neighborhoods chosen have a substantial influence on the algorithm's performance, it is very important to select the neighborhood structures appropriate to the problem structure.

Algorithm: Variable Neighborhood Search

Begin

(1) N_k, Choosing neighborhood structure sets ($k = 1, \ldots, k_{max}$)
(2) x, Finding the initial solution
(3) Determining stopping condition

<u>**Solution Procedure**</u>

Continue following steps until meeting the stopping condition

(1) Set k as 1
(2) Repeat following steps until $k = k_{max}$
- (a) <u>Shaking</u>: Generate a solution (x') from *k'th* Neighborhood structure of
  $(x' : N_k(x))$
- (b) <u>Local Search</u>: Apply local search method to the initial solution x' $(x'' : N_m(x'))$. If the solution found in this step is best found solution set as x''.
- (c) <u>Move or Not</u>: If the local optimum found is better than incumbent, sent $x = x''$ and continue with same neighborhood. Otherwise, set $k = k + 1$

End

Fig. 1. Variable neighborhood search algorithm

Todosijević et al. (2017) used 2-opt and OR-opt moves as they are two of the most common neighborhood structures implemented to solve TSP. Mjirda et al. (2017) analyzed the effects of sequential use of neighborhoods. Within this scope, on the one hand, the effect of 2-Opt, insertion-1 and insertion-2 neighboring structures on the algorithm performance was examined and on the other hand the effects of greedy heuristic and random permutation methods were analyzed. As a result of the computational experiments, it was reported that the initial solution, neighborhood structure and order of neighborhood used in the algorithm had a significant role on the performance of the solution approach. Soylu (2015), in solving multiple TSP, developed a general VNS algorithm utilizing Or-opt move, Two-point move, Three-point move, 2-opt move utilizing them sequentially and confirmed the effect of each operator on the performance of the algorithm.

Pseudo-code of proposed algorithm is presented in Fig. 2.

Procedure: VNS approach
Input: parameters, structures, a_{max} , b_{max} …
Output: x
Read data for problem instance
Initialize algorithm parameters
Select $N_a, a = 1, \ldots, a_{max}$ as operators for shaking phase
Select $N_b, b = 1, \ldots, b_{max}$ as operators for local search phase
Find initial solution x^0 randomly and apply feasibility procedure
Set VNS termination condition
$x \leftarrow x^0$
While iteration number$<$maximum number of iterations **do**
  Set $a \leftarrow 1$
  While $a < a_{max}$ **do**
    $x' \leftarrow$shake with shaking structures $((N_a(x))$ -*Randomly selected*
    While b$<$ b_{max} **do**
      $x'' \leftarrow$ best local search with local search structures $((N_a(x'))$ -*Randomly selected*
      $x \leftarrow x''$
      $a \leftarrow 1$
      If $f(x) < f(x^0)$ **then**
        $x^0 \leftarrow x$
        $a \leftarrow 1$
        $b \leftarrow 1$
      Else
        $b \leftarrow b + 1$
      End if
    End while
    $a \leftarrow a + 1$
  End while
  If iteration limit for VNS is exceeded optionally apply perturbation
End while

Fig. 2. Pseudo-code of proposed algorithm

For the proposed algorithm, neighborhood structures which are presented in Table 3 have been used:

Table 3. Neighborhood structures used in the proposed algorithm

Phase	Operator	Description
Shaking	Random insertion	A node is selected randomly and inserted at a random place in the same route
	Random feasible insert	A node is selected randomly and inserted at a random position if and only if generated route is feasible
	Random Feasible 2-Opt	Deleting 2 different edges randomly and adding new edges while fulfilling feasibility conditions and maintaining tour structure
Local search	Best insertion	The best insertion movement that removes a client from its position and inserts to another position in the same route
	Best 2-Opt	The best two-opt movement that swaps pairs of links in the same route is applied to the route
	Best Block Insert	A block of nodes selected and inserted at a place in the same route

While for the shaking phase, random insertion (Kalayci et al. 2016), random feasible insert and random feasible 2-opt operators were used in a random sequence (Polat et al. 2015), best of insertion, best of 2-opt and best block insertion were used in local search step. Along with the use of operators randomly, concerning best block insert operator (Kalayci and Kaya 2016) used for local search, number of nodes inside of the blocks were also determined randomly up to 3 nodes.

Along with the neighborhood structure to be used, initial solution to be obtained has a vital importance in terms of the solution performance. Within the scope of this study, initial solution has been generated randomly and if necessary, feasibility correction has been applied afterwards.

4 Computational Experiments

4.1 Implementation

The proposed approach has been modelled and tested using MATLAB R2017a and numerical experiments were run on an Intel Xeon E5-2650 2.0 GHz processor with 32 GB RAM.

In the following subsections, a sensitivity analysis on the datasets and computational results for each test problem are presented.

4.2 Computational Results

Proposed algorithm has been tested on twenty-two different publicly available data sets. These instances contain 10 to 900 nodes with different edge costs. In order to analyze and evaluate algorithm performance on different sized networks, datasets have been categorized as small datasets containing 10–100 nodes, medium datasets containing 150–400 nodes and large datasets containing 500–900 nodes. Brief summary of datasets presented in Table 4.

Table 4. Summary of test instances

No	Scale	Number of nodes	Number of instances
1	Small	10–100	12
2	Medium	150–400	5
3	Large	500–900	5

The experimental study was carried out in two stages by executing the algorithm for $100 \times number\ of\ Nodes$ iterations. In the first step, the VNS algorithm developed for this problem was tested on data sets without perturbation mechanism and the results were collected. In the second stage, the algorithm was supported with perturbation mechanism and the effect of this relevant mechanism on the algorithm performance and results was observed. The results obtained by proposed solution approach has been presented in Table 5. Ten replications have been performed for each datasets and worst, average and best results (with perturbation/without perturbation) as well as average CPU time have been collected.

When the results are examined, the contribution of the perturbation mechanism to the algorithm performance is clearly seen. Although this effect is less noticeable in small sized problem instances, the greater gap between best results found is much more evident as seen in Fig. 3.

Moreover, as also mentioned in previous sections, increase of the problem dimension greatly affects the execution time of the algorithm. The CPU time greatly increases with respect to the problem size as it is clearly seen in the Fig. 4.

5 Conclusion and Discussion

In this study, one of the recent variants of TSP, cost-balanced TSP, was solved by a VNS algorithm. Within this framework, this study was limited to a basic implementation of the variable neighborhood search algorithm that systematically explores the solution space with simple insert and 2-opt operators. For the shaking step of the algorithm, random insertion, random feasible insert and 2-Opt operators were used and best of insertion, best of 2-Opt and best block insertion operators were used for local search phase. Perturbation mechanism was optionally applied to the algorithm and its effect

Table 5. Results obtained by proposed algorithm

Number of nodes	Result				Average CPU time in secs
	Worst	Average	Best without perturbation	Best with perturbation	
10	105	105	105	**105**	0.2
15	283	273	271	**271**	1.1
20	339	309	296	**296**	1.6
25	410	397	386	**375**	3.4
30	500	479	433	**433**	8.9
40	625	555	514	**458**	13.3
50	861	767	717	**630**	14.1
60	1.041	985	922	**833**	21.4
70	1.157	1.095	1.011	**868**	29.8
80	1.192	1.127	1.047	**862**	44.1
90	1.462	1.364	1.271	**1.029**	54.5
100	1.588	1.433	1.272	**1.030**	68.8
150	2.895	2.680	2.513	**1.793**	117.8
200	3.552	3.340	3.154	**2.349**	352.7
250	4.837	4.558	4.375	**3.273**	499.4
300	6.428	6.081	5.868	**4.602**	735.8
400	8.578	8.375	8.074	**6.241**	1358.9
500	10.725	10.358	9.844	**8.351**	3687.4
600	12.944	12.455	12.019	**10.321**	5572.1
700	15.908	15.470	14.896	**12.360**	6565.5
800	18.976	17.911	17.214	**15.896**	7209.3
900	20.332	19.472	18.813	**16.796**	8461.2

on the solution performance was observed. Applying perturbation mechanism has a significant effect on the large-sized instances.

The proposed solution approach is very simple to implement, and the performance is acceptable. However, several improvements can be made on the current version of the algorithm to enhance exploration and exploitation capabilities via different neighborhood structures and local search algorithms. Quality of the obtained initial solution may also have a vital importance on the algorithm's performance. Therefore, rather than beginning with random feasible initial solution, obtaining better initial solution would undoubtfully decrease the execution time of the algorithm. Although shaking procedure in the variable neighborhood search algorithm tries to escape from local optima, a perturbation mechanism is necessary for a more efficient exploration of the search space.

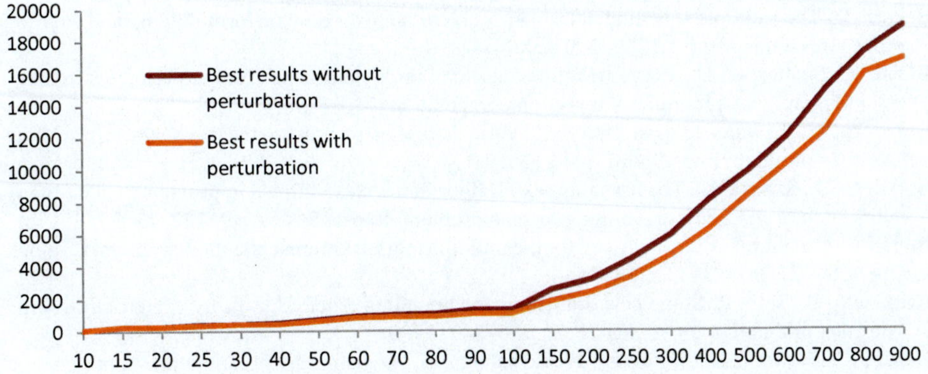

Fig. 3. Comparison of best results with/without perturbation mechanism

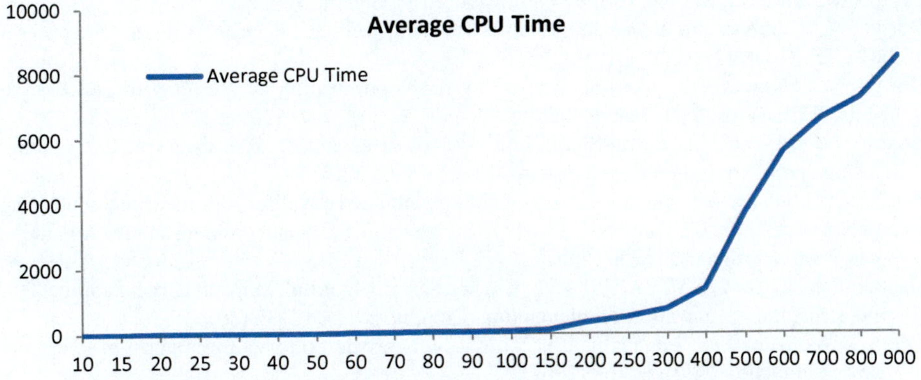

Fig. 4. Increase of the CPU time according to the problem dimension

Moreover, how often to apply this mechanism is worthy of investigation that would require an extensive amount of experiments.

Acknowledgement. This study was supported by the Scientific Research Coordination Unit of Pamukkale University under the project number: 2019KRM004-013.

References

Akhand, M.A.H., Akter, S., Rashid, M.A.: Velocity tentative particle swarm optimization to solve TSP. In: 2013 International Conference on Electrical Information and Communication Technology (EICT), 13–15 February 2014, pp. 1–6 (2014)

Applegate, D., Bixby, R., Chvátal, V., Cook, W.: Implementing the Dantzig-Fulkerson-Johnson algorithm for large traveling salesman problems. Math. Program. **97**(1), 91–153 (2003)

Balas, E., Martin, G.: ROLL-A-ROUND: software package for scheduling the rounds of a rolling mill. Copyright Balas and Martin Associates **104** (1985)

Bektas, T.: The multiple traveling salesman problem: an overview of formulations and solution procedures. Omega **34**(3), 209–219 (2006)
Bland, R., Shallcross, D.: Large traveling salesman problem arising from experiments in x-ray crystallography. A preliminary report on computation (730) (1987)
Boryczka, U., Szwarc, K.: An effective hybrid harmony search for the asymmetric travelling salesman problem. Eng. Optim. 1–17 (2019a)
Boryczka, U., Szwarc, K.: The harmony search algorithm with additional improvement of harmony memory for asymmetric traveling salesman problem. Expert Syst. Appl. **122**, 43–53 (2019b)
Butt, S.E., Cavalier, T.M.: A heuristic for the multiple tour maximum collection problem. Comput. Oper. Res. **21**(1), 101–111 (1994)
Carpaneto, G., Toth, P.: Some new branching and bounding criteria for the asymmetric travelling salesman problem. Manag. Sci. **26**(7), 736–743 (1980)
Dahan, F., El Hindi, K., Mathkour, H., AlSalman, H.: Dynamic flying ant colony optimization (DFACO) for solving the traveling salesman problem. Sensors **19**(8), 1837 (2019)
Dantzig, G., Fulkerson, R., Johnson, S.: Solution of a large-scale traveling-salesman problem. J. Oper. Res. Soc. Am. **2**(4), 393–410 (1954)
Dorigo, M., Birattari, M.: Ant Colony Optimization. Springer (2010)
Dorigo, M., Gambardella, L.M.: Ant colonies for the travelling salesman problem. Biosystems **43**(2), 73–81 (1997)
Dorigo, M., Maniezzo, V., Colorni, A.: Ant system: optimization by a colony of cooperating agents. IEEE Trans. Syst. Man Cybern. B Cybern. **26**(1), 29–41 (1996)
Escario, J.B., Jimenez, J.F., Giron-Sierra, J.M.: Ant colony extended: experiments on the travelling salesman problem. Expert Syst. Appl. **42**(1), 390–410 (2015)
Fischetti, M., Salazar-Gonzalez, J.-J., Toth, P.: The generalized traveling salesman and orienteering problems. In: Gutin, G., Punnen, A.P. (eds.) The Traveling Salesman Problem and Its Variations, pp. 609–662. Springer, Boston (2007)
Gandomi, A.H., Yang, X.-S., Alavi, A.H.: Cuckoo search algorithm: a metaheuristic approach to solve structural optimization problems. Eng. Comput. **29**(1), 17–35 (2013)
Geem, Z.W., Kim, J.H., Loganathan, G.V.: A new heuristic optimization algorithm: harmony search. Simulation **76**(2), 60–68 (2001)
Gendreau, M., Laporte, G., Semet, F.: The covering tour problem. Oper. Res. **45**(4), 568–576 (1997)
Grötschel, M., Jünger, M., Reinelt, G.: Optimal control of plotting and drilling machines: a case study. Zeitschrift für Oper. Res. **35**(1), 61–84 (1991)
Hansen, P., Mladenovic, N.: Variable neighborhood search: principles and applications. Eur. J. Oper. Res. **130**(3), 449–467 (2001)
Holland, J.H.: Adaptation in Natural and Artificial Systems. The University of Michigan Press, Ann Arbor (1975)
Hore, S., Chatterjee, A., Dewanji, A.: Improving variable neighborhood search to solve the traveling salesman problem. Appl. Soft Comput. **68**, 83–91 (2018)
Ismkhan, H.: Effective heuristics for ant colony optimization to handle large-scale problems. Swarm Evol. Comput. **32**, 140–149 (2017)
Jünger, M., Reinelt, G., Rinaldi, G.: The traveling salesman problem, Chapter 4. In: Handbooks in Operations Research and Management Science, vol. 7, pp. 225–330. Elsevier (1995)
Kalayci, C.B., Kaya, C.: An ant colony system empowered variable neighborhood search algorithm for the vehicle routing problem with simultaneous pickup and delivery. Expert Syst. Appl. **66**, 163–175 (2016)
Kalayci, C.B., Polat, O., Gupta, S.M.: A hybrid genetic algorithm for sequence-dependent disassembly line balancing problem. Ann. Oper. Res. **242**(2), 321–354 (2016)
Karaboga: An idea based on honey bee swarm for numerical optimization. Technical report-tr06, Erciyes University, Engineering Faculty, Computer Engineering Department (2005)

Karp, R.M.: Reducibility among combinatorial problems. In: Miller, R.E., Thatcher, J.W., Bohlinger, J.D. (eds.) Complexity of Computer Computations: Proceedings of a symposium on the Complexity of Computer Computations, held 20–22 March 1972, at the IBM Thomas J. Watson Research Center, Yorktown Heights, New York, and sponsored by the Office of Naval Research, Mathematics Program, IBM World Trade Corporation, and the IBM Research Mathematical Sciences Department, pp. 85–103. Springer, Boston (1972)

Kennedy, J., Eberhart, R.: Particle swarm optimization. In: Proceedings of the IEEE International Conference on Neural Networks, November/December 1995, vol. 4, pp. 1942–1948 (1995)

Kinable, J., Smeulders, B., Delcour, E., Spieksma, F.C.R.: Exact algorithms for the equitable traveling salesman problem. Eur. J. Oper. Res. **261**(2), 475–485 (2017)

Kobeaga, G., Merino, M., Lozano, J.A.: An efficient evolutionary algorithm for the orienteering problem. Comput. Oper. Res. **90**, 42–59 (2018)

Kocer, H.E., Akca, M.R.: An improved artificial bee colony algorithm with local search for traveling salesman problem. Cybern. Syst. **45**(8), 635–649 (2014)

Labbé, M., Laporte, G., Martín, I.R., González, J.J.S.: The median cycle problem (1999)

Lahyani, R., Khemakhem, M., Semet, F.: A unified matheuristic for solving multi-constrained traveling salesman problems with profits. EURO J. Comput. Optim. **5**(3), 393–422 (2017)

Laporte, G.: The traveling salesman problem: an overview of exact and approximate algorithms. Eur. J. Oper. Res. **59**(2), 231–247 (1992)

Laporte, G., Martello, S.: The selective travelling salesman problem. Discret. Appl. Math. **26**(2), 193–207 (1990)

Lenstra, J.K., Kan, A.H.G.R.: Some simple applications of the travelling salesman problem. J. Oper. Res. Soc. **26**(4), 717–733 (1975)

Lourenço, H.R., Martin, O.C., Stützle, T.: Iterated local search. In: Handbook of Metaheuristics, pp. 320–353. Springer (2003)

Mavrovouniotis, M., Müller, F.M., Yang, S.: Ant colony optimization with local search for dynamic traveling salesman problems. IEEE Trans. Cybern. **47**(7), 1743–1756 (2017)

Mehrabian, A.R., Lucas, C.: A novel numerical optimization algorithm inspired from weed colonization. Ecol. Inform. **1**(4), 355–366 (2006)

Meng, X., Li, J., Dai, X., Dou, J.: Variable neighborhood search for a colored traveling salesman problem. IEEE Trans. Intell. Transp. Syst. **19**(4), 1018–1026 (2017)

Menger, K.: Das botenproblem. Ergebnisse eines mathematischen kolloquiums **2**, 11–12 (1932)

Mjirda, A., Todosijević, R., Hanafi, S., Hansen, P., Mladenović, N.: Sequential variable neighborhood descent variants: an empirical study on the traveling salesman problem. Int. Trans. Oper. Res. **24**(3), 615–633 (2017)

Mladenovic, N., Hansen, P.: Variable neighborhood search. Comput. Oper. Res. **24**(11), 1097–1100 (1997)

Osaba, E., Yang, X.-S., Diaz, F., Lopez-Garcia, P., Carballedo, R.: An improved discrete bat algorithm for symmetric and asymmetric Traveling Salesman Problems. Eng. Appl. Artif. Intell. **48**, 59–71 (2016)

Ouaarab, A., Ahiod, B., Yang, X.-S.: Discrete cuckoo search algorithm for the travelling salesman problem. Neural Comput. Appl. **24**(7), 1659–1669 (2014)

Palomo-Martínez, P.J., Salazar-Aguilar, M.A.: The bi-objective traveling purchaser problem with deliveries. Eur. J. Oper. Res. **273**(2), 608–622 (2019)

Plante, R.D., Lowe, T.J., Chandrasekaran, R.: The product matrix traveling salesman problem: an application and solution heuristic. Oper. Res. **35**(5), 772–783 (1987)

Polat, O., Kalayci, C.B., Kulak, O., Gunther, H.O.: A perturbation based variable neighborhood search heuristic for solving the vehicle routing problem with simultaneous pickup and delivery with time limit. Eur. J. Oper. Res. **242**(2), 369–382 (2015)

Ramesh, T.: Traveling purchaser problem. Opsearch **18**(1–3), 78–91 (1981)

Ratliff, H.D., Rosenthal, A.S.: Order-picking in a rectangular warehouse: a solvable case of the traveling salesman problem. Oper. Res. **31**(3), 507–521 (1983)

Saji, Y., Riffi, M.E.: A novel discrete bat algorithm for solving the travelling salesman problem. Neural Comput. Appl. **27**(7), 1853–1866 (2016)

Sevkli, Z., Sevilgen, F.E.: Variable neighborhood search for the orienteering problem. In: Computer and Information Sciences – ISCIS 2006, pp. 134–143. Springer, Heidelberg (2006)

Shokouhifar, M., Jalali, A., Torfehnejad, H.: Optimal routing in traveling salesman problem using artificial bee colony and simulated annealing. In: 1st National Road ITS Congress (2015)

Soylu, B.: A general variable neighborhood search heuristic for multiple traveling salesmen problem. Comput. Ind. Eng. **90**, 390–401 (2015)

Tang, L., Wang, X.: An iterated local search heuristic for the capacitated prize-collecting travelling salesman problem. J. Oper. Res. Soc. **59**(5), 590–599 (2008)

Todosijević, R., Mjirda, A., Mladenović, M., Hanafi, S., Gendron, B.: A general variable neighborhood search variants for the travelling salesman problem with draft limits. Optim. Lett. **11**(6), 1047–1056 (2014)

Wang, J., Ersoy, O.K., He, M., Wang, F.: Multi-offspring genetic algorithm and its application to the traveling salesman problem. Appl. Soft Comput. **43**, 415–423 (2016)

Wang, Y., Chen, Y., Lin, Y.: Memetic algorithm based on sequential variable neighborhood descent for the minmax multiple traveling salesman problem. Comput. Ind. Eng. **106**, 105–122 (2017)

Yang, X.-S.: A new metaheuristic bat-inspired algorithm. In: González, J.R., Pelta, D.A., Cruz, C., Terrazas, G., Krasnogor, N. (eds.) Nature Inspired Cooperative Strategies for Optimization (NICSO 2010), pp. 65–74. Springer, Heidelberg (2010)

Zhong, Y., Lin, J., Wang, L., Zhang, H.: Hybrid discrete artificial bee colony algorithm with threshold acceptance criterion for traveling salesman problem. Inf. Sci. **421**, 70–84 (2017)

Zhou, Y., Luo, Q., Chen, H., He, A., Wu, J.: A discrete invasive weed optimization algorithm for solving traveling salesman problem. Neurocomputing **151**, 1227–1236 (2015)

Adaptive Iterated Local Search with Random Restarts for the Balanced Travelling Salesman Problem

Jacopo Pierotti[1](✉), Lorenzo Ferretti[2], Laura Pozzi[2], and J. Theresia van Essen[1]

[1] Delft Institute of Applied Mathematics, Delft University of Technology, Delft, The Netherlands
{j.pierotti,j.t.vanessen}@tudelft.nl
[2] Faculty of Informatics, Università della Svizzera italiana, Lugano, Switzerland
{lorenzo.ferretti,laura.pozzi}@usi.ch

Abstract. Metaheuristics have been widely used to solve NP-hard problems, with excellent results. Among all NP-hard problems, the Travelling Salesman Problem (TSP) is potentially the most studied one. In this work, a variation of the TSP is considered; the main differences being, edges may have positive or negative costs and the objective is to return a Hamiltonian cycle with cost as close as possible to zero. This variation is called the *balanced* TSP (BTSP). To tackle this new problem, we present an adaptive variant of the iterated local search metaheuristic featuring also random restart. This algorithm was tested on the MESS2018 metaheuristic competition and achieved notable results, scoring the 5th position overall. In this paper, we detail all the components of the algorithm itself and present the best solutions identified. Even though this metaheuristic was tailored for the BTSP, with small modifications its structure can be applied to virtually any NP-hard problem. In particular, we introduce the *uneven reward-and-punishment* rule which is a powerful tool, applicable in many contexts where fast responses to dynamic changes are crucial.

Keywords: Iterated Local Search · Travelling Salesman Problem · Balanced Travelling Salesman Problem · Hamiltonian cycle · Metaheuristic

1 Introduction

In the Travelling Salesman Problem (TSP), a salesman has to visit a given set of cities and, after travelling along all of them, has to return to the one he started from, hence, completing a cycle. Given a set of cities V and a cost matrix D with entries $d_{i,j}$ $(i, j \in V)$ –in the *standard* TSP, costs represent the distance between cities–, the goal of the traveller is to find the cycle of minimum cost covering all the cities. The TSP is an NP-hard problem, and therefore, there are no

S. Greco et al. (Eds.): MESS 2018, AISC 1332, pp. 37–56, 2021.
https://doi.org/10.1007/978-3-030-68520-1_4

known techniques able to provide an optimal solution efficiently. In this work, we consider a variation of the *standard* TSP. In particular, we address the *balanced* TSP (BTSP). In this variant, the entries of the cost matrix can be negative and the goal of the problem is to find the cycle, visiting all the cities, with cost as close as possible to zero. Despite the complexity of these problems, the TSP and its variants are widely studied and used in many different contexts. Typical applications are: clustering [1], vehicle routing [2], computer wiring [3], cutting wallpaper [4], job-scheduling [5], DNA sequencing [6] and pattern-allocation [7] problems. To cope with these problems and their complexity, many approaches have been proposed in the literature. The main discriminant among them is their ability to obtain an optimal solution or to rapidly discover a near-optimal one. In the latter, there is a trade off between the time required to identify the solution and the quality of the solution provided.

Among the different approaches proposed in literature, some of the most relevant are: Genetic Algorithms (GA), Ant Colony Optimisation (ACO), Tabu Search (TS), Adaptive Large Neighbourhood Search (ALNS), Simulating Annealing (SA), Local Search (LS) and Iterated Local Search (ILS). The approach proposed by Juneja et al. [8] exploits the ability of population-based heuristics to search for multiple solutions in each iteration of the algorithm and, by using various combinations of selection, crossover and mutation techniques, to continuously improve the quality of current solutions. Dorigo et al. [9] were the first to introduce the possibility to use ACO, a heuristic algorithm which navigates the solution space by mimicking ants finding food as a group, as a viable strategy to solve the TSP. More recently, Escario et al. [10] have refined this approach introducing different types of agents –specialised ants– and population dynamics to organise the ants' movements. The TS approach proposed by Toth et al. [11] is based on the use of restricted neighbourhoods, allowing to reduce the solution space and leading to a more efficient implementation of candidate strategies proposed for tabu search algorithms. The ALNS heuristic, proposed by Ribeiro et al. [12], is based on the algorithm initially devised by Ropke and Pisinger [13] and extends the large neighbourhood search of Shaw [14] by using destroy and repair methods within the same search process. Differently from the previous metaheuristics, SA techniques mimic the metal annealing process by considering probabilistic moves depending on a temperature parameter. The SA methodologies proposed in [15] and [16] have been adopted for the TSP resulting in a viable alternative to the above mentioned metaheuristics. Lastly, LS and ILS techniques have been vastly used to solve the TSP, such as in [17,18] and [19]. As described by Lourenço et al. in [20], by using LS, a sequence of viable solutions is iteratively generated within the embedded heuristic.

In this work, we propose a variation of the ILS technique. ILS, differently from LS techniques, alternates the search phase with the perturbation phase in order to escape tenacious local optima. Given a graph, our approach searches for a Hamiltonian cycle within it and, by iterating over a sequence of actions operating on the current solution, navigates the neighbourhood of the current solution while searching for local optima. Perturbations are then applied in order to escape local optima, hence searching different regions of the solution space. The main contribution of this work consists in the introduction of an *uneven reward-and-punishment* adaptation rule, which in turn leads to a more reactive response to the current solution.

The paper continues as follows: Sect. 2 presents the problem formulation, Sect. 3 describes the advantages of ILS techniques and introduces as proposed methodology an Adaptive Iterated Local Search with Random Restarts (AILS-RR). Section 4 shows experimental evidence for the effectiveness of the AILS-RR technique introduced, and finally, Sect. 5 concludes the paper and presents final remarks.

2 Problem Formulation

The goal of the TSP is to minimise the cost of the cycle connecting a set of given cities that a salesman has to visit exactly once. The cycle of the salesman can be defined as a sub-graph of graph $G(V, E)$ where V is the set of vertices –cities– and E is the set of edges of the graph representing the connections between cities. Given two cities $i, j \in V$, we define $d_{\overline{ij}}$ as the cost of travelling from city i to city j. In addition, we define a binary variable $x_{\overline{ij}}$ which specifies if the cycle of the salesman includes the edge from city i to city j as follows:

$$x_{\overline{ij}} = \begin{cases} 1, & \text{if edge } \overline{ij} \in E \text{ is in the cycle of the salesman} \\ 0, & \text{otherwise.} \end{cases} \tag{1}$$

Then, we define the total cost of the cycle as:

$$C_{total} = \sum_{\overline{ij} \in E} d_{\overline{ij}} x_{\overline{ij}}. \tag{2}$$

In the *classic* TSP, all costs are strictly positive and the objective is to find the Hamiltonian cycle of minimum cost. In the BTSP, costs can be either positive or negative and the objective is to find the Hamiltonian cycle of cost as close as possible to zero. A Hamiltonian cycle is a connected sequence of edges that joins a sequence of vertices, such that each vertex in V is visited exactly once and the edge sequence is closed. A closed sequence starts from a vertex i and, through an ordered sequence of vertices connected by edges, returns to the original vertex i. Given an ordered sequence, we name adjacent vertices of a generic vertex its previous and its following vertex. In general, graphs are not fully connected, i.e. not all vertices $i, j \in V$ are connected by a direct edge. Edges are undirected and each of them is associated with a cost which can be either positive or negative. We define the degree of a vertex as the number of its outgoing edges.

The differences between the BTSP and the TSP translate the model of the *classic* TSP in a similar one, hereby reported in (3)–(8):

$$\min C_{total} \tag{3}$$

$$C_{total} \geq \sum_{\overline{ij} \in E} d_{\overline{ij}} x_{\overline{ij}} \tag{4}$$

$$C_{total} \geq - \sum_{\overline{ij} \in E} d_{\overline{ij}} x_{i,j} \tag{5}$$

$$\sum_{j \in V} x_{\overline{ij}} = 2 \qquad \forall i \in V \tag{6}$$

$$\sum_{i,j \in T, i \neq j} x_{\overline{ij}} \leq |T| - 1 \qquad \forall T \subset V, T \neq \emptyset \tag{7}$$

$$x_{\overline{ij}} \in \{0, 1\} \qquad \forall \overline{ij} \in E \tag{8}$$

$$C_{tot} \in \mathbb{R}^{+} \tag{9}$$

Objective function (3) and Constraints (6), (7) and (8) are standard TSP constraints. In particular, Constraints (6) impose the cycle to visit each vertex exactly once –one incoming, one outgoing edge– while Constraints (7) avoid the presence of sub-tours T in the cycle. Constraints (8) force variables $x_{\overline{ij}}$ to be binary and Objective function (3) aims at minimising the absolute cost of the cycle. Lastly, considering that is a minimisation problem, Constraints (4), (5) and (9) impose variable C_{tot} to assume the absolute value of the cost of the cycle. By minimising the absolute value, we force the costs to be as close to zero as possible.

To solve this problem, we used AILS-RR, which is variant of ILS. We define as a feasible solution any Hamiltonian cycle and as objective function the cost of the cycle itself. Thus, we aim at identifying the best candidate in the neighbourhood of the current solution by applying modifications to the solution structure.

3 Methodology

The Iterated Local Search methodology repeatedly builds a sequence of solutions generated by a local search heuristic embedded in a framework. As defined by Lourenço et al. in [20], given a current solution s, ILS generates an intermediate solution $\hat{s}$ by applying changes on s. Then, the embedded local search heuristic is applied to $\hat{s}$, which leads to a new solution s'. If this solution improves s, it becomes the new current solution in our sequence, and thus we keep navigating the solution space modifying s'. Otherwise, if s' does not introduce an improvement with respect to s, ILS continues to apply modification to s. Starting from a feasible solution, ILS explores its neighbourhood and determines the best solution within it. Thus, this metaheuristic exploits the possibility to search for a solution in a reduced space defined by the output of the local search heuristic, instead of searching over the entire solution space.

The methodology proposed in this paper, i.e. AILS-RR, relies on ILS by searching in the neighbourhood of an already existing solution. In order to apply AILS-RR, it is necessary to generate a first solution to be fed to the AILS-RR

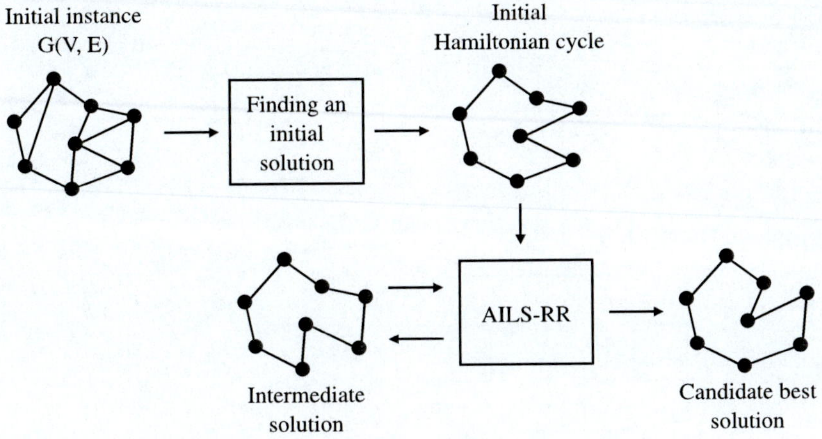

Fig. 1. The AILS-RR framework and its phases.

itself. Our methodology starts generating an initial solution from a graph $G(V, E)$ representing a BTSP instance. Once a solution has been identified, this is given as input to AILS-RR, which iteratively searches for improving solutions until a certain stopping criterion is reached. Figure 1 shows an overview of the steps of our methodology.

Details about the strategy adopted to identify an initial solution are presented in Sect. 3.1, while the AILS-RR procedure is described in Sect. 3.2.

3.1 Finding an Initial Solution

Determining whether there exists a Hamiltonian cycle in a not fully connected graph is an NP-complete problem [21]. In order to find a Hamiltonian cycle, we used the Snakes-and-Ladders Heuristics (SLH) by Baniasadi et al. [22]. The SLH is a polynomial time algorithm inspired by the k-opt heuristic. In SLH, vertices are ordered on a circle where edges between adjacent vertices represent the arcs of the circle while all other edges are considered as chords of the circle. The heuristic attempts to place all edges of a Hamiltonian cycle on the circle by seeking changes in the arrangement of vertices of the graph, with the goal of maximising the number of edges on the circle.

For this work, we have relied on the online implementation of SLH [23] provided by the authors of [22]. However, the online implementation accepts a maximum of 2000 vertices. To generate a cycle in the instances with more than 2000 vertices, we randomly partitioned all the vertices in equally sized subsets, with size smaller than 2000. For each of them, we independently found a Hamiltonian cycle; then, considering only edges from one subset to another, we selected the k vertices with highest degree. Finally, for each of the k vertices and its adjacent ones in the Hamiltonian cycle, namely A and B, we checked if they

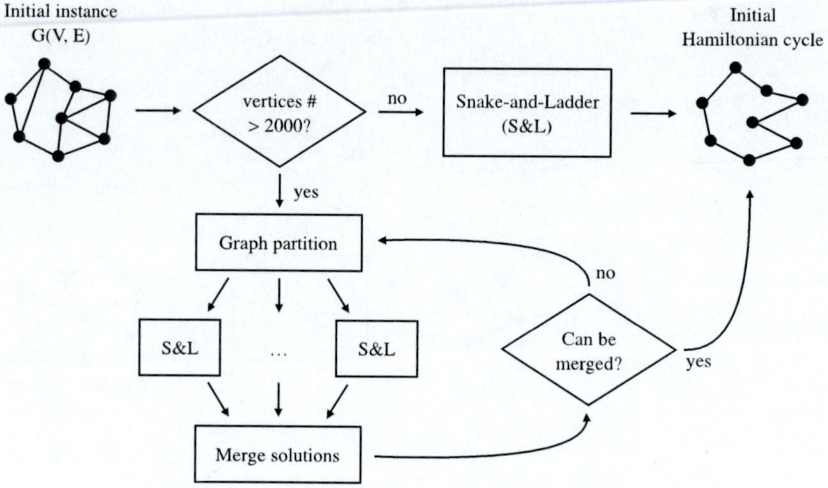

Fig. 2. Flowchart of the process to generate the initial solution.

were connected to any couple of adjacent vertices in the other subset, namely C and D. If so, it means that the two disjoint cycles can be united as $A-$*first cycle*$-B-C-$*second cycle*$-D-A$. When none of the k vertices could be used to join the two cycles, the whole procedure was repeated by randomly partitioning the nodes. Figure 2 shows a flowchart of the process used to identify an initial solution. For all the evaluated instances, we were able to find an initial solution without having to re-partition more than two times.

3.2 Adaptive Iterated Local Search with Random Restarts

Once an initial feasible solution is found, we used AILS-RR to improve its objective value. The AILS-RR procedure proposed in this work relies on four main phases: *local search*, *update*, *perturbation* and *random restart*.

The procedure starts from the initial solution s provided by the method described in Sect. 3.1. Then, a *local search* is performed on s and a solution s' is returned. This solution can be: a) a new solution different from s or b) the same solution s in case no improving solution has been identified. Thus, the *update* phase amends the local search heuristic according to the resulting solution generated. In this *update* phase, the adaptive part of the algorithm takes place.

Once solution s' has been generated and the *update* has been performed, we replace s_{best} – the best solution found so far – with s' if the latter strictly improves it. In case no improved solution, with respect to s, was found after $MaxIteration$ consecutive iterations, a *perturbation* to s is applied and a new solution is generated. Then, the next AILS-RR iteration starts from the perturbed solution and the counter for the not improving solution is reset. Additionally, if for *too many* consecutive iterations, no solution has improved s_{best}, a *random restart* from a *good* known solution is performed. In particular, details on the number of consecutive iterations needed and on what is considered a *good* solution are presented in Sect. 3.2.4. Random restart is applied to avoid extensively searching unpromising regions of the solution space, hence moving to a more fruitful one. Finally, the AILS-RR terminates whenever the stopping criterion is met. Further details on the stopping criterion are given in Sect. 3.2.5.

A pseudo-code of the approach is illustrated in Algorithm 1. In the algorithm, there are five main functions: a) *LocalSearch*, b) *Update*, c) *Perturbation*, d) *RandomRestart* and e) *StoppingCriterion*. Our algorithm differentiates from the *standard* ILS [20] in terms of the update phase as well as the introduction of the random restart from a known solution. In the following paragraphs, we explain each component of the algorithm in detail.

Algorithm 1. Adaptive Iterated Local Search with Random Restart

```
procedure AILS (Input: Graph G, Hamiltonian cycle s; Output: Hamiltonian
  cycle s_best)
    s_best = s;
    notImproving = i = 0;
    while not StopCriterion() do
        i = 0;
        while i < MaxIterations do
            s'=LocalSearch(s);
            Update(LocalSearch());
            if Cost(s') < Cost(s) then
                s = s';
                i = 0;
                if |Cost(s')| < |Cost(s_best)| then
                    s_best = s';
                    notImproving = 0;
                else
                    notImproving++;
                end if
            else
                i++; notImproving++;
                if notImproving > RestartFactor then
                    s = RandomRestart();
                    notImproving = 0;
                end if
            end if
        end while
        s=Perturbation(s);
    end while
    return s_best
end procedure
```

3.2.1 Local Search

Starting from a current solution s, the local search aims at finding an improved solution s'. It consists of applying modifications, which are dictated by different operators, to the solution structure. An operator is a function that, given cycle s, applies modifications on its structure which generates multiple cycles which are variations of cycle s. The resulting cycles define a neighbourhood of s. Every solution in the neighbourhood is evaluated and only the solution which most improves s is accepted. If no solution improves s, s itself is returned. During the local search, the algorithm uses – with probability depending on their weights – one of these three operators: *one edge insertion*, *two edges insertion* and *cycle modification*. Each operator selects at least one edge to be inserted in s. The insertion of an edge divides the original cycle in two subtours. Selecting the edge to be inserted determines which subtours will be created. Dually, identifying a desired subtour lets us establish which edge is to be inserted. Given the particular structure of the instances, we have chosen to determine the edges first. More on this is presented in Sect. 4.1. The following paragraphs introduce the operators adopted.

One Edge Insertion. Given a cycle s, the first operator selects, at random, one edge $\overline{AB}$ which is not in s. The extreme points of the edge, A and B, are adjacent to two vertices each in s –C and D for A, E and F for B. For the time being, we assume every vertex to be different with respect to each other; straightforward modifications can be applied if this is not the case. There is only a limited number of possibilities to insert edge $\overline{AB}$ in the existing solution; at most, there are eight possible outcomes. The cost, Eq. (3), of all possible outcomes is evaluated and the best one is chosen. Since the graph is not complete, in general not all the combinations exist. Indeed, naming p the probability that an edge exists, and assuming they are all independent, we can analyse quantitatively the probability for each combination to exist. Figure 3 shows all possible outcomes; the edge to be inserted is represented in blue, the edges that may or may not exist are shown in red, while black indicates the edges belonging to the original cycle. By construction, we know the existence of edges $\overline{AB}, \overline{AC}, \overline{AD}, \overline{BE}$ and $\overline{BF}$. Hence, we deduce there are two combinations with probability p, four combinations with probability p^2 and two combinations with probability p^3.

Two Edges Insertion. Similarly to the previous operator, this process chooses two edges not yet in the current solution and tries to insert them. If the four extreme vertices of the two selected edges are all different, isolating them divides the cycle in four subtours. Hence, the solution is now decomposed in six subtours – four from the original cycle and two from the two inserted edges, that can be considered subtours as well, see Fig. 4ii.

There are 10!![1] possible ways to combine the four subtours and the two edges. This number comes from $(2\cdot(t-1))!!$, where t is the number of subtours – six, in our case – and -1 because a degree of freedom is lost for the intrinsic symmetry

[1] !! is double factorial, i.e. $f!! = f\cdot(f-2)\cdot(f-4)...$
In our case, $10!! = 3840$.

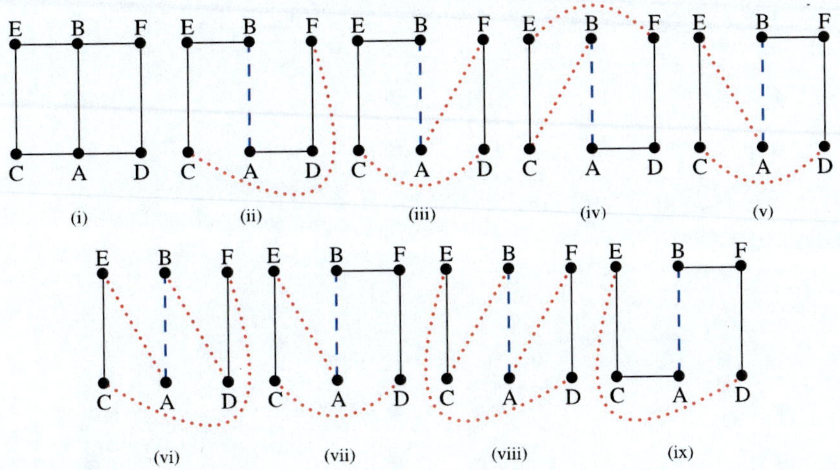

Fig. 3. Examples of single edge insertion. Dashed blue lines show the edge to be inserted; dotted red lines indicate the edges that may or may not exist, while black lines show the edges belonging to the original cycle. Figure (i) shows the original cycle, while figures (ii)–(ix) show the possible insertions.

of cycles. A multiplicative factor of 2 is added since each subtour can be linked to the next one through two different endpoints. Having t subtours implies having, as their endpoints, $2t$ vertices. Intuitively, the double factorial follows because a vertex can be connected to $2t - 2$ other vertices, every vertex but itself and the other endpoint of its subtour. Once connected, the following vertex can be connected to $2t - 4$ others. This includes all the vertices but itself, the other endpoint of its subtour and the endpoints of the subtour to which it is already linked to. Recursively, we can see how this develops, for the remaining vertices, in a double factorial structure. Among these combinations, only the ones with at least probability p^3 to exist are considered by our methodology.

Generally speaking, these first two operators can be viewed as modified versions of k-opt. Figure 4 shows an example of two edges insertion. Starting from an initial cycle – Fig. 4i – two edges are inserted. The new edges divide the cycle into four different subtours – Fig. 4ii. Finally, Fig. 4iii and Fig. 4iv show an example of a reconstructed cycle with probability p^3 and p^6, respectively.

Cycle Modification. The two edge insertion generates multiple intermediate solutions, but it is computationally more expensive with respect to the one edge insertion operator. To compensate the computational requirements of the two edge insertion operator, we introduce the cycle modification operator.

This operator selects, at random, an edge in the existing solution s. We name A and B its extreme vertices, which are consecutive in the original solution s. Then, we select at random one edge, not in solution s, which is outgoing A and is entering, without loss of generality, in C. At the same time, we select at random one edge, not in solution s, which is outgoing B and is entering,

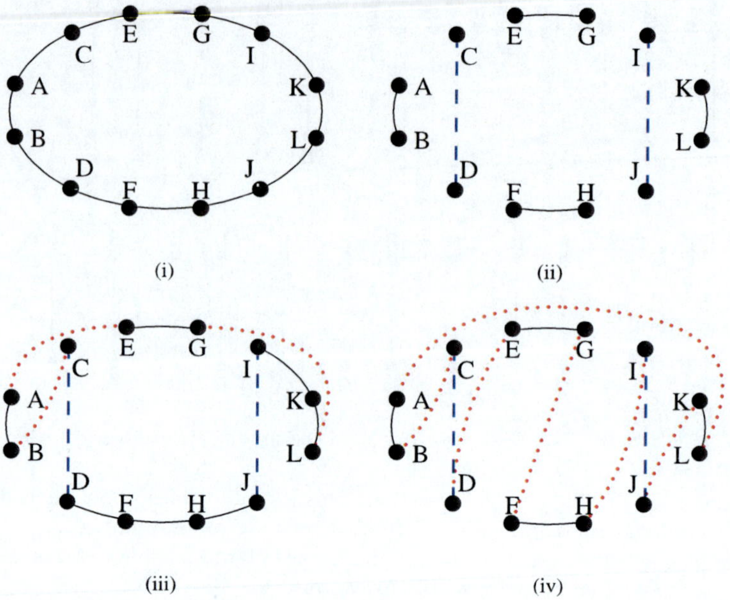

Fig. 4. Example of two edge modification. (i) Initial cycle. (ii) Insertion of two edges and subtours generated. (iii) Reconstructed cycle with probability p^3. (iv) Reconstructed cycle with probability p^6. The original edges are shown in black, while dashed blue lines depict the inserted edges, and dotted red lines depict the edges with probability p.

without loss of generality, in $D \neq C$, see Fig. 5ii. Subsequently, we consider the path, in the original cycle, from C to D, that passes through A and B. In that cycle, we name E and F the follower of C and the predecessor of D, respectively. By construction, there exist paths $\overline{EA}, \overline{CD}, \overline{BF}$ and edges $\overline{AC}, \overline{BD}$. Hence, there exists a path connecting $\overline{EA} - \overline{AC} - \overline{CD} - \overline{DB} - \overline{BF}$, see Fig. 5iii. Finally, if edge $\overline{EF}$ exists, we obtain a feasible cycle, see Fig. 5iv. In general, edge $\overline{EF}$ exists with probability p. To increase the size of the neighbourhood, this procedure is repeated for all outgoing edges of B. It is not, however, repeated for all combinations of outgoing edges of A and outgoing edges of B, because this would be computationally too expensive.

3.2.2 Update

Each operator of the local search is applied with a probability proportional to its associated weight. These weights are constrained to be greater than a parameter $MinWeight$ and their sum is forced to a value lower that the upperbound parameter $MaxWeights$. Whenever an operator returns a solution which does not improve the input solution, we subtract f –in this work, f has value 1– from its associated weight. In case an operator returns a better solution than the solution given as input, its associated weight is increased by 10% and rounded

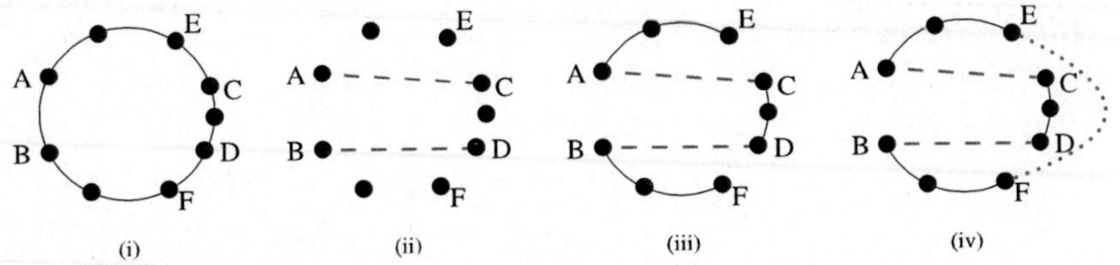

Fig. 5. Example of cycle modification. (i) Original cycle. (ii) Selection of the edges to be inserted. (iii) Construction of the existing path. (iv) Closing the cycle with edge $\overline{EF}$ which exists with probability p. Original edges are shown in black, dashed blue lines depict the inserted edges, and dotted red lines indicate the edges that exist with probability p.

to the nearest higher integer. In addition, if the returned solution is even better than the best known solution $-s_{best}-$ the weight of the operator leading to the improved solution receives an extra reward of $10f$ in addition to the normal reward obtained for improving the previous solution. We call this discrepancy among a constant decrease and a proportional increase an *uneven reward-and-punishment* adaptation rule. In our opinion, an even reward-and-punishment adaptation rule is more suited to grasp stable characteristics, such as the ones related to the structure of the *graph* itself, while an uneven rule is more keen to quickly adapt to variations, such as the ones in the changing structure of the *solution*.

3.2.3 Perturbation

The *perturbation* is applied when we are not able to improve a local solution for a significant number of iterations $-MaxIterations$. To perform a perturbation, the algorithm uses the same operators as the local search. The main differences, with respect to the local search, is that every change is accepted –not only an improving one– and it is performed only once. There is no evidence that more perturbations results in better solutions. In fact, more perturbations cause the current solution to drift too much away from a promising part of the solution space. In addition, since the costs of the edges are neither Euclidean nor the authors found any pattern within them, even a slight modification of a few edges can lead to dramatic changes in the objective function.

3.2.4 Random Restart

Perturbations allow to explore different regions of the solution space; nonetheless, some of those regions could be unpromising. To avoid exploring inadequate regions of the solution space, it is useful to restart the search from a region where *good* solutions are known to exist. If, after *too many* consecutive iterations, no solution improved the best known objective function, then a *random restart* from a *good* known solution is performed. In particular, we define as $History$ an array storing the $HistorySize$ best solutions and their number of occurrences. If, after

RestartFactor consecutive not improving iterations, s_{best} was not improved, we perform a random restart from any of the solutions stored in *History*.

We define *RestartFactor* as:

$$RestartFactor = cMax + \frac{MostVistitedSolution(History)}{HistoryStep}, \tag{10}$$

where $MostVistitedSolution(History)$ assumes the value of the number of visits to the most visited solution in *History*, while *cMax* and *HistoryStep* are parameters. *cMax* indicates the minimum number of iterations the algorithm has to perform before a *random restart* can happen, while *HistoryStep* is a scaling factor. While we perform the random restart to avoid going too far away from a region of the solution space where good solutions exist, we reduce the frequency of restarts when the same solution is visited more and more times to escape that tenacious local minimum. In fact, it could happen that too frequent restarts drives the local search to the same local minima. In addition, restarting from any of the solutions stored in *History* helps to maintain a certain degree of diversity.

3.2.5 Stopping Criterion

AILS-RR has no memory of all the solutions discovered since it started, and in general there is no guarantee of optimality. Hence, without a stopping criterion, it would indefinitely search for improving solutions. The stopping criterion we implemented terminates the execution of the algorithm if any of the following conditions is met: a) the solution cost is zero, and thus, we have reached the optimal solution, b) a user-defined time limit was exceeded, c) the algorithm returned for more than *MaxIterationHistory* times the same solution.

If condition a) is met, then, it is not possible to further improve the solution identified. Condition b) offers a knob for setting a reasonable usage of resources required to search for improving solutions, and condition c) is useful to avoid expensive explorations of particularly tenacious local minima from where the algorithm cannot escape even with its *perturbation* move.

4 Experiments

In this section, we explain the experiments setup and the performance of our algorithm. In Sect. 4.1, we describe the instances tested, in Sect. 4.2, the parameters used, and lastly, in Sect. 4.3, the performance of our algorithm.

4.1 Instances

The algorithm was tested on 27 given instances, available at [24], which vary in size from 10 vertices and 40 edges, to 3,000 vertices and 12,000 edges. Hence, on average, each vertex has degree 4. This motivates the analysis on the probability of existence of an edge in the AILS-RR. The absolute value of the costs of every edge can be written as $k_1 \cdot 100{,}000 + k_2$, where k_1 and k_2 are integers in the range $[0, 99]$. We run the algorithm twice per instance. The first time, we used as input the instances considering as cost only k_1. In the following, we refer to this change in the cost associated with the edges as a cost modification. With this data, the algorithm was able to find a solution of cost zero for all instances. Then, the algorithm was run a second time, starting from the previously found solution, with the real costs of the arcs.

4.2 Parameters Tuning

The local search procedure is repeated until for $MaxIterations = 100$ consecutive iterations no improving solution is found. $HistorySize$, the number of how many *good* solutions were stored in array $History$, is set to 100. Parameters $cMax$ and $HistoryStep$, which are used to determine when to restart from a random solution in $History$, are set to 1,000 and 100, respectively. Parameter $MaxIterationHistory$ determines how many times a solution can be visited before the stopping criterion is met and is set to 1,000,000. This means that a random restart can happen as often as after 1,000 consecutive not improving iterations, or as rarely as after 10,999 consecutive not improving iterations. In Paragraph 3.2.4, we explained how often the random restart happens depending on how many times the most inspected solution is visited. We may have a restart after 10,999 consecutive not improving iterations and not after 11,000 times as expected if $MostVisitedSolution(History)$ assumes value $MaxIterationHistory$. In fact, $MostVisitedSolution(History)$ cannot assume value $MaxIterationHistory$ in Eq. (10) because, if so, the stopping criterion is met and the execution of the whole algorithm is terminated. Every single operator weight is initially set to 333 and restricted to integer values above $MinWeight = 1$ and such that their sum does not exceed $MaxWeights = 1{,}000$. If the sum of the weights exceeds $MaxWeights$, the weight of every parameter is decreased by one, unless this violates the lower bound $MinWeight$, until the threshold is respected. For the tests with the modified cost, the maximum running time for each instance was set to two hours while, for the tests with the original cost, the maximum running time for each instance was set to twelve hours. Table 1 summarises all the parameters used in the algorithm.

4.3 Performance

Instances were run overnight on different machines. In particular, instances up to 100 nodes were run on an Intel Core i7-6600U CPU @2.60 GHz 2.80 GHz with 8 GB RAM and instances from 150 to 400 nodes on a Intel Core i7 @2.9 GHz with 8 GB RAM. Bigger instances (500 to 3,000 nodes) were run on a 32 core machine with Intel Xeon E5-4650L CPU @2.60 GHz 3.1 GHz with 500 GB of physical memory. Since the BTSP is a new problem, introduced for the MESS2018 solver challenge, no comparison with the state of the art is possible. In general, optimal solutions are not known but they cannot have a better objective function than zero. By executing the AILS-RR on the instances with modified costs of the

Table 1. Parameters tuning

Parameter name	Value	Description
MaxIterations	100	Maximum not improving cycles of LS
HistorySize	100	Dimension of array *History*
cMax	1,000	Minimum number of cycles for a random restart
HistoryStep	100	Random restart parameter
MaxIterationHistory	1,000,000	Stopping condition parameter
MinWeight	1	Minimum weight for each operator
MaxWeights	1,000	Maximum value of the sum of the weights of all operators
MaxTime	2 h	Maximum time per instance - modified costs
MaxTime	12 h	Maximum time per instance - original costs

edges, as explained in Sect. 4.1, we know that the optimal solutions for all these modified instances have cost zero. In general, a zero cost solution for the modified instances does not translate into an optimal solution for the original instances; nonetheless, it is a *good* initial when solving the instance with original costs. The results displayed in Sect. 4.3.1 and 4.3.2 refer to the modified costs and the results shown in Sect. 4.3.3 refer to the original costs. For the tests with the modified costs, we set a time limit of two hours and a limit of 10,000 iterations by counting how many times *LocalSearch* is called. We ran this experiment to perform a qualitative analysis of the obtained results and operator effectiveness. We tested the instances on a 32 core machine with Intel Xeon E5-4650L CPU @2.60 GHz 3.1 GHz with 500 GB of physical memory, and each test was run 10 times in order to obtain average results. In Sect. 4.3.1, we introduce in detail a meaningful instance case, while in Sect. 4.3.2, we present results for all instances.

4.3.1 Instance 3,000 Vertices

The instance presented in this section is representative of the entire set. In fact, Fig. 6i and Fig. 6ii display the trends of the objective function, for the same executions, with respect to iterations and time while Fig. 6iii and Fig. 6iv show the evolution of the weights during the ten tests of the algorithm, with respect to iterations and time. Since all the tests are plotted simultaneously, these figures give some idea on the variance of the trends and how many tests terminated their execution in just a few iterations. First of all, plotting results with respect to iterations or with respect to time only slightly modifies the overall shape of the figures. This is due to the fact that comparable amounts of time are needed for each operator to perform its local search. In Fig. 6iii and Fig. 6iv, sharp peaks with slow decline are visible. This is exactly the effect of the *uneven reward-and-punishment adaptation rule*; since increases are proportional while decreases are constant, rapid changes in the weights of the operators are visible. In this case, it is clear that operator *cycle modification* performs better than the others; as shown in Sect. 4.3.2, this is the case for basically all other instances. Secondly, Fig. 6i and Fig. 6ii show the absolute value of the best solution found so far by the algorithm. Even though this instance is the biggest one, even for the worst of the ten tests, our algorithm was able to find an optimal solution in roughly two minutes.

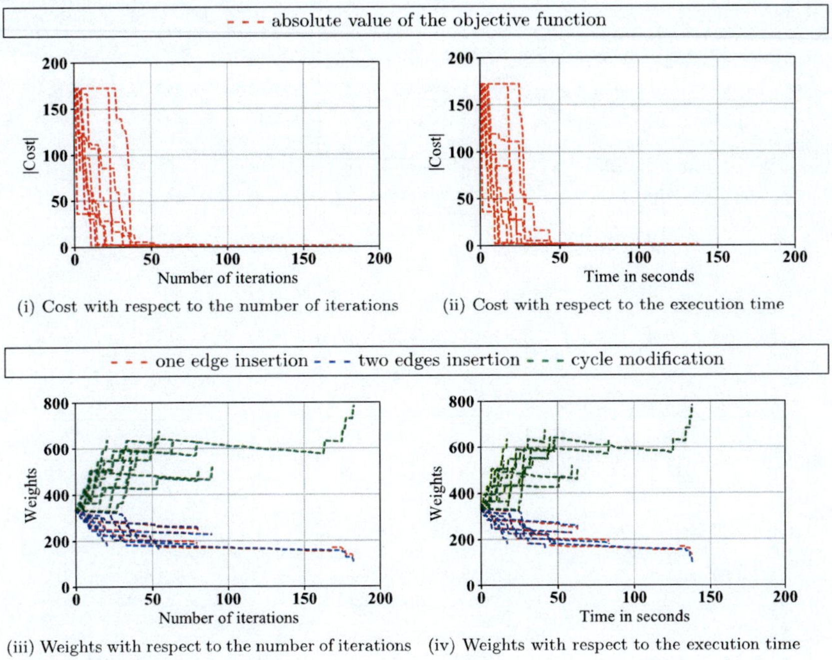

Fig. 6. Evolution of absolute cost of the solutions and operator weights with respect to number of iterations and execution time. Results are shown for the instance with 3,000 vertices.

4.3.2 Results for All Instances

Small instances were solved in a few iterations with no particularly interesting trend; because of that, in this paragraph, we consider only instances of size strictly greater than one hundred vertices. Since no particular difference arises from plotting with respect to the number of iterations or with respect to time, the figures in this paragraph refer to the iterations. Furthermore, for the sake of readability, we decided to plot average results instead of all the 10 trends. Averaging the results highlights the trends but smooths peaks which instead are visible in Fig. 6iii and Fig. 6iv. For instances with more than one hundred nodes, trends of the weights of the operators and of the solution developments are shown in Fig. 7 and Fig. 8, respectively. These trends show the average results for the ten tests. Figure 7 shows that, for all simulations, all tests over all the instances, but one, returned the optimal solution within few iterations –resulting

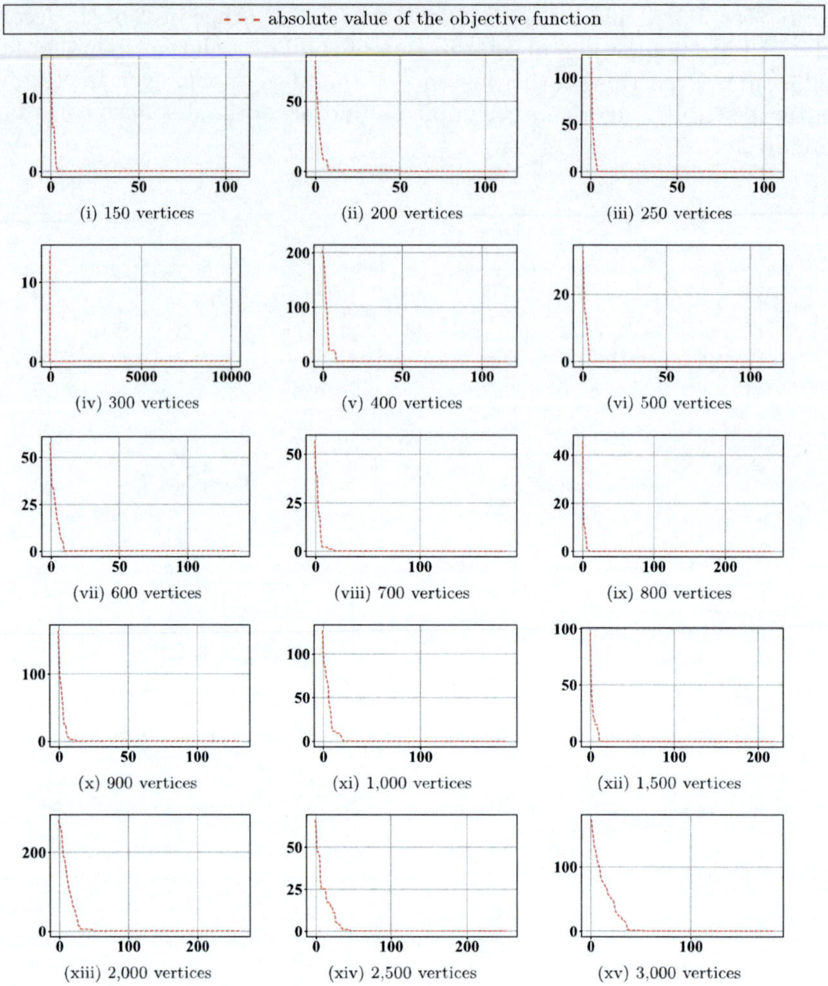

Fig. 7. Evolution of the objective function over different instances. Number of iterations on the x-axis, objective function value on the y-axis.

in few minutes of execution time–, way before encountering the time or the iteration limit. In our opinion, this is a powerful indicator of the effectiveness of our algorithm. Similarly, Fig. 8, shows how among all the instances, the *cycle modification* is the most effective operator. Nonetheless, it is worth noticing that, while for the medium-sized instances, weights are *almost* equivalently distributed among operators, the bigger the instance, the greater the probability of *cycle modification* to be chosen.

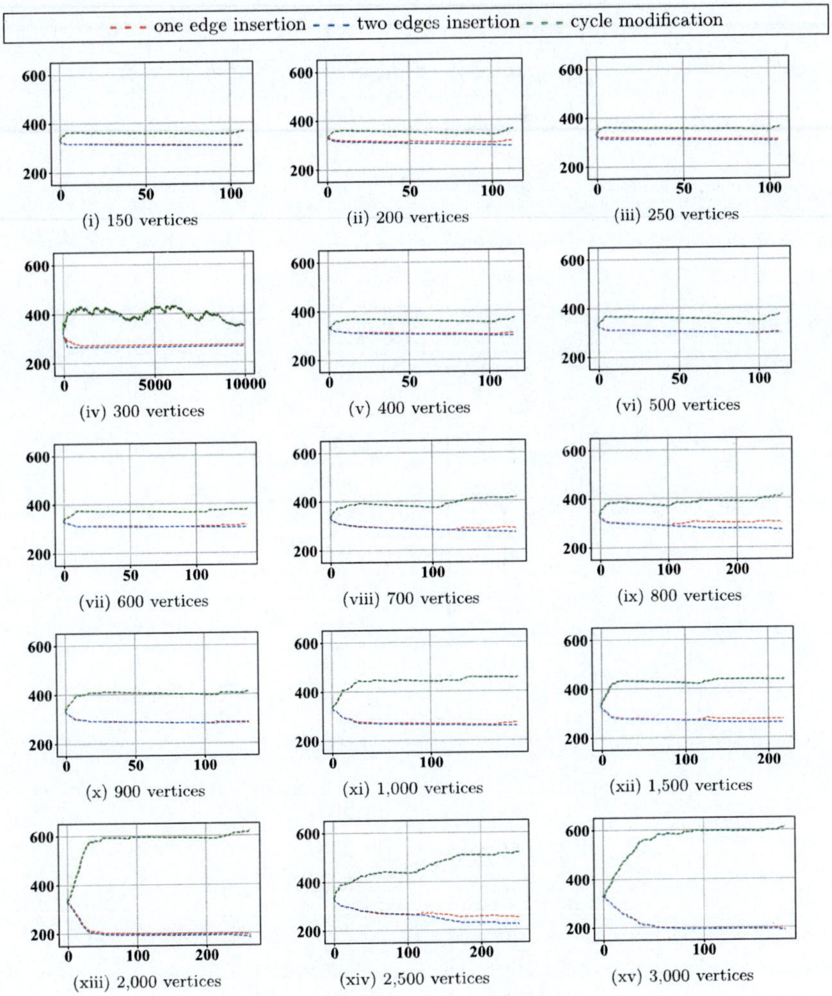

Fig. 8. Evolution of the weights assigned to the different operators over different instances. Number of iterations on the x-axis, objective function value on the y-axis.

4.3.3 Final Results

In this section, for the sake of further comparison, we display the results submitted to the competition. All the results proposed in this section are computed with the original costs. In particular, Table 2 portraits: in the first column, the instance size –expressed in the number of vertices–, and in the second column, the absolute value of the solutions.

Table 2. Results.

# vertices	|Solution cost|
10	105
15	271
20	296
25	375
30	433
40	473
50	717
60	918
70	1,056
80	929
90	1,098
100	1,245
150	2,035
200	2,657
250	3,811
300	4,846
400	6,509
500	8,418
600	9,784
700	17,989
800	18,233
900	20,596
1,000	22,597
1,500	37,662
2,000	49,882
2,500	36,607
3,000	24,423

5 Conclusion

This paper illustrates the AILS-RR methodology applied to the balances travelling salesman problem. With slight modifications of the local search operators, we believe that the same metaheuristic can obtain significant results in many operational research problems. The proposed metaheuristic is a variant of ILS and it features the adaptive use of the local search operators and restart moves. Key advantages of the AILS-RR are: its effectiveness in navigating the solution space, as shown in the achieved ranking in the MESS2018 Metaheuristics Competition, its easiness to implement and its ability to quickly obtain near optimal solutions. Additional motivations and a detailed description of our algorithm are presented in Sect. 3, which presents the algorithm structure focusing on the different phases of the metaheuristic. In particular, the description details the main contribution of the proposed methodology which lays in the newly introduced *uneven reward-and-punishment adaptation* rule. To the best of our knowledge, this is the first time that such a strategy is used. Section 4 proves that our AILS-RR achieves notable results, scoring remarkable positions in almost every instance ranking, and achieving the 5th position in the competition.

Acknowledgement. The authors want to thank the organisers of the MESS2018 summer school for the challenging opportunity they offered us. The first author wants to acknowledge the DIAMANT mathematics cluster for partially funding his work.

References

1. Lenstra, J.K.: Clustering a data array and the traveling-salesman problem. Oper. Res. **22**(2), 413–414 (1974)
2. Lenstra, J.K., Kan, A.R.: Complexity of vehicle routing and scheduling problems. Networks **11**(2), 221–227 (1981)
3. Lenstra, J.K., Kan, A.R.: Some simple applications of the travelling salesman problem. J. Oper. Res. Soc. **26**(4), 717–733 (1975)
4. Hahsler, M., Hornik, K.: TSP-infrastructure for the traveling salesperson problem. J. Stat. Softw. **23**(2), 1–21 (2007)
5. Whitley, L.D., Starkweather, T., Fuquay, D.: Scheduling problems and traveling salesmen: the genetic edge recombination operator. In: 3rd International Conference on Genetic Algorithms, vol. 89, pp. 133–140 (1989)
6. Caserta, M., Voß, S.: A hybrid algorithm for the DNA sequencing problem. Discret. Appl. Math. **163**, 87–99 (2014)
7. Madsen, O.B.: An application of travelling-salesman routines to solve pattern-allocation problems in the glass industry. J. Oper. Res. Soc. **39**(3), 249–256 (1988)
8. Juneja, S.S., Saraswat, P., Singh, K., Sharma, J., Majumdar, R., Chowdhary, S.: Travelling salesman problem optimization using genetic algorithm. In: 2019 Amity International Conference on Artificial Intelligence, pp. 264–268. IEEE (2019)
9. Dorigo, M., Gambardella, L.M.: Ant colonies for the travelling salesman problem. Biosystems **43**(2), 73–81 (1997)
10. Escario, J.B., Jimenez, J.F., Giron-Sierra, J.M.: Ant colony extended: experiments on the travelling salesman problem. Expert Syst. Appl. **42**(1), 390–410 (2015)
11. Toth, P., Vigo, D.: The granular tabu search and its application to the vehicle-routing problem. Inf. J. Comput. **15**(4), 333–346 (2003)
12. Ribeiro, G.M., Laporte, G.: An adaptive large neighborhood search heuristic for the cumulative capacitated vehicle routing problem. Comput. Oper. Res. **39**(3), 728–735 (2012)
13. Ropke, S., Pisinger, D.: An adaptive large neighborhood search heuristic for the pickup and delivery problem with time windows. Transp. Sci. **40**(4), 455–472 (2006)
14. Shaw, P.: A new local search algorithm providing high quality solutions to vehicle routing problems. APES Group, Department of Computer Science, University of Strathclyde, Glasgow, Scotland, UK (1997)

15. Geng, X., Chen, Z., Yang, W., Shi, D., Zhao, K.: Solving the traveling salesman problem based on an adaptive simulated annealing algorithm with greedy search. Appl. Soft Comput. **11**(4), 3680–3689 (2011)
16. Malek, M., Guruswamy, M., Pandya, M., Owens, H.: Serial and parallel simulated annealing and tabu search algorithms for the traveling salesman problem. Ann. Oper. Res. **21**(1), 59–84 (1989)
17. Johnson, D.S.: Local optimization and the traveling salesman problem. In: International Colloquium on Automata, Languages, and Programming, pp. 446–461. Springer (1990)
18. Voudouris, C., Tsang, E.: Guided local search and its application to the traveling salesman problem. Eur. J. Oper. Res. **113**(2), 469–499 (1999)
19. Paquete, L., Stützle, T.: A two-phase local search for the biobjective traveling salesman problem. In: International Conference on Evolutionary Multi-Criterion Optimization, pp. 479–493. Springer (2003)
20. Lourenço, H.R., Martin, O.C., Stützle, T.: Iterated local search. In: Handbook of Metaheuristics, pp. 320–353. Springer (2003)
21. Garey, M., Johnson, D., Tarjan, R.: The planar Hamiltonian circuit problem is NP-complete. SIAM J. Comput. **5**(4), 704–714 (1976)
22. Baniasadi, P., Ejov, V., Filar, J.A., Haythorpe, M., Rossomakhine, S.: Deterministic "Snakes and Ladders" Heuristic for the Hamiltonian cycle problem. Math. Program. Comput. **6**(1), 55–75 (2014)
23. Snake and Ladders Heuristic. http://www.flinders.edu.au/science_engineering/csem/research/programs/flinders-hamiltonian-cycle-project/slhweb-interface.cfm. Accessed 01 Aug 2018
24. MESS Competition. https://195.201.24.233/mess2018/home.html. Accessed 01 Aug 2018

Correction to: Metaheuristics for Combinatorial Optimization

Salvatore Greco, Mario F. Pavone, El-Ghazali Talbi,
and Daniele Vigo

Correction to:
S. Greco et al. (Eds.): *Metaheuristics for Combinatorial Optimization*, AISC 1332,
https://doi.org/10.1007/978-3-030-68520-1

The book was inadvertently published with the incorrect volume number (1336) and this has been corrected now (1332).

The updated version of the book can be found at
https://doi.org/10.1007/978-3-030-68520-1

S. Greco et al. (Eds.): MESS 2018, AISC 1332, p. C1, 2021.
https://doi.org/10.1007/978-3-030-68520-1_5

Author Index

S. Greco et al. (Eds.): MESS 2018, AISC 1332, p. 57, 2021.
https://doi.org/10.1007/978-3-030-68520-1